ARTIFICIAL INTELLIGENCE AND DATA DRIVEN OPTIMIZATION OF INTERNAL COMBUSTION ENGINES

ARTIFICIAL INTELLIGENCE AND DATA DRIVEN OPTIMIZATION OF INTERNAL COMBUSTION ENGINES

Edited by

JIHAD BADRA

Transport Technologies Division, Research and Development Center, Saudi Aramco, Dhahran, Eastern Province, Saudi Arabia

PINAKI PAL

Energy Systems Division, Argonne National Laboratory, Lemont, IL, United States

YUANJIANG PEI

Aramco Americas: Aramco Research Center—Detroit, MI, United States

SIBENDU SOM

Energy Systems Division, Argonne National Laboratory, Lemont, IL, United States

ELSEVIER

Elsevier
Radarweg 29, PO Box 211, 1000 AE Amsterdam, Netherlands
The Boulevard, Langford Lane, Kidlington, Oxford OX5 1GB, United Kingdom
50 Hampshire Street, 5th Floor, Cambridge, MA 02139, United States

Copyright © 2022 Elsevier INC. All rights reserved. UChicago Argonne, LLC, Contract No: DE-AC02-06CH11357.

No part of this publication may be reproduced or transmitted in any form or by any means, electronic or mechanical, including photocopying, recording, or any information storage and retrieval system, without permission in writing from the publisher. Details on how to seek permission, further information about the Publisher's permissions policies and our arrangements with organizations such as the Copyright Clearance Center and the Copyright Licensing Agency, can be found at our website: www.elsevier.com/permissions.

This book and the individual contributions contained in it are protected under copyright by the Publisher (other than as may be noted herein).

Notices
Knowledge and best practice in this field are constantly changing. As new research and experience broaden our understanding, changes in research methods, professional practices, or medical treatment may become necessary.

Practitioners and researchers must always rely on their own experience and knowledge in evaluating and using any information, methods, compounds, or experiments described herein. In using such information or methods they should be mindful of their own safety and the safety of others, including parties for whom they have a professional responsibility.

To the fullest extent of the law, neither the Publisher nor the authors, contributors, or editors, assume any liability for any injury and/or damage to persons or property as a matter of products liability, negligence or otherwise, or from any use or operation of any methods, products, instructions, or ideas contained in the material herein.

Library of Congress Cataloging-in-Publication Data
A catalog record for this book is available from the Library of Congress

British Library Cataloguing-in-Publication Data
A catalogue record for this book is available from the British Library

ISBN: 978-0-323-88457-0

For information on all Elsevier publications visit our website at https://www.elsevier.com/books-and-journals

Publisher: Matthew Deans
Acquisitions Editor: Carrie Bolger
Editorial Project Manager: Mariana Kuhl
Production Project Manager: Poulouse Joseph
Cover Designer: Greg Harris

Typeset by TNQ Technologies

Contents

Contributors ix
Foreword xi
Preface xv

1. **Introduction** 1
   Balaji Mohan, Pinaki Pal, Jihad Badra, Yuanjiang Pei and Sibendu Som
   1. Industrial revolution 1
   2. Artificial intelligence, machine learning, and deep learning 2
   3. Machine learning algorithms 3
   4. Artificial intelligence—based fuel-engine co-optimization 4
   5. Summary 16
   References 16

SECTION 1 Artificial Intelligence to optimize fuel formulation

2. **Optimization of fuel formulation using adaptive learning and artificial intelligence** 27
   Juliane Mueller, Namho Kim, Simon Lapointe, Matthew J. McNenly, Magnus Sjöberg and Russell Whitesides
   1. Introduction and motivation 27
   2. Mixed-mode combustion and fuel performance metrics 28
   3. A neural network model to predict fuel research octane numbers 31
   4. Optimization problem formulation and description of solution approaches 32
   5. Numerical experiments and results 37
   6. Discussion 40
   7. Summary and concluding remarks 42
   Acknowledgments 43
   References 43

3. **Artificial intelligence—enabled fuel design** 47
   Kiran K. Yalamanchi, Andre Nicolle and S. Mani Sarathy
   1. Transportation fuels 47
   2. Application of artificial intelligence to fuel formulation 52
   3. Conclusions and perspectives 58
   Acknowledgments 60
   References 60

SECTION 2 Artificial Intelligence and computational fluid dynamics to optimize internal combustion engines

4. Engine optimization using computational fluid dynamics and genetic algorithms — 71

Alberto Broatch, Ricardo Novella, José M. Pastor, Josep Gomez-Soriano and Peter Kelly Senecal

1. Introduction — 71
2. Modeling framework and acceleration strategies — 74
3. Optimization methods — 79
4. Summary and concluding remarks — 97
References — 98

5. Computational fluid dynamics—guided engine combustion system design optimization using design of experiments — 103

Yuanjiang Pei, Anqi Zhang, Pinaki Pal, Le Zhao, Yu Zhang and Sibendu Som

1. Introduction — 103
2. Methodologies — 106
3. A recent application — 113
4. Recommendations for best practice — 118
5. Conclusions and perspectives — 120
Acknowledgments — 121
References — 121

6. A machine learning-genetic algorithm approach for rapid optimization of internal combustion engines — 125

Jihad Badra, Opeoluwa Owoyele, Pinaki Pal and Sibendu Som

1. Introduction — 125
2. Engine optimization problem setup — 127
3. Training and data examination — 129
4. Machine learning-genetic algorithm approach — 132
5. Automated machine learning-genetic algorithm — 141
6. Summary — 156
Acknowledgments — 156
References — 156

7. **Machine learning—driven sequential optimization using dynamic exploration and exploitation** — 159

Opeoluwa Owoyele and Pinaki Pal

1. Introduction — 159
2. Active ML optimization (ActivO) — 160
3. Case study 1: two-dimensional cosine mixture function — 165
4. Case study 2: computational fluid dynamics (CFD)-based engine optimization — 171
5. Conclusions — 179
Acknowledgments — 180
References — 180

SECTION 3 Artificial Intelligence to predict abnormal engine phenomena

8. **Artificial-intelligence-based prediction and control of combustion instabilities in spark-ignition engines** — 185

Bryan Maldonado, Anna Stefanopoulou and Brian Kaul

1. Introduction — 185
2. Case study: artificial-intelligence-enhanced modeling of dilute spark-ignition cycle-to-cycle variability — 189
3. Case study: neural networks for combustion stability control — 193
4. Case study: learning reference governor for model-free dilute limit identification and avoidance — 199
5. Summary — 204
References — 205

9. **Using deep learning to diagnose preignition in turbocharged spark-ignited engines** — 213

Eshan Singh, Nursulu Kuzhagaliyeva and S. Mani Sarathy

1. Introduction — 213
2. Preignition detection using machine learning algorithm — 215
3. Activation functions — 221
4. Experiments and data extraction — 222
5. Machine learning methodology — 224
6. Model 1: Input from principal component analysis — 230
7. Model 2: Time series input — 231

viii   Contents

    8. Model metrics   232
    9. Results and discussion   233
   10. Conclusions   234
   References   235
   Further reading   236

Index   *239*

Contributors

Jihad Badra
Transport Technologies Division, Research and Development Center, Saudi Aramco, Dhahran, Eastern Province, Saudi Arabia

Alberto Broatch
CMT—Thermal Motors, Polytechnic University of Valencia, Camino de Vera, Valencia, Spain

Josep Gomez-Soriano
CMT—Thermal Motors, Polytechnic University of Valencia, Camino de Vera, Valencia, Spain

Brian Kaul
Oak Ridge National Laboratory, Knoxville, TN, United States

Namho Kim
Sandia National Laboratories, Livermore, CA, United States

Nursulu Kuzhagaliyeva
Clean Combustion Research Center (CCRC), King Abdullah University of Science and Technology, Thuwal, Western Province, Saudi Arabia

Simon Lapointe
Lawrence Livermore National Laboratory, Livermore, CA, United States

Bryan Maldonado
Oak Ridge National Laboratory, Knoxville, TN, United States

Matthew J. McNenly
Lawrence Livermore National Laboratory, Livermore, CA, United States

Balaji Mohan
Transport Technologies Division, Research and Development Center, Saudi Aramco, Dhahran, Eastern Province, Saudi Arabia

Juliane Mueller
Lawrence Berkeley National Laboratory, Berkeley, CA, United States

Andre Nicolle
Aramco Fuel Research Center, Aramco Overseas, Rueil-Malmaison, Paris, France

Ricardo Novella
CMT—Thermal Motors, Polytechnic University of Valencia, Camino de Vera, Valencia, Spain

Opeoluwa Owoyele
Energy Systems Division, Argonne National Laboratory, Lemont, IL, United States

Pinaki Pal
Energy Systems Division, Argonne National Laboratory, Lemont, IL, United States

José M. Pastor
CMT—Thermal Motors, Polytechnic University of Valencia, Camino de Vera, Valencia, Spain

Yuanjiang Pei
Aramco Americas: Aramco Research Center—Detroit, Novi, MI, United States

S. Mani Sarathy
Clean Combustion Research Center (CCRC), King Abdullah University of Science and Technology, Thuwal, Western Province, Saudi Arabia

Peter Kelly Senecal
Convergent Science, Inc., Madison, WI, United States

Eshan Singh
Clean Combustion Research Center (CCRC), King Abdullah University of Science and Technology, Thuwal, Western Province, Saudi Arabia

Magnus Sjöberg
Sandia National Laboratories, Livermore, CA, United States

Sibendu Som
Energy Systems Division, Argonne National Laboratory, Lemont, IL, United States

Anna Stefanopoulou
Department of Mechanical Engineering, University of Michigan, Ann Arbor, MI, United States

Russell Whitesides
Lawrence Livermore National Laboratory, Livermore, CA, United States

Kiran K. Yalamanchi
Clean Combustion Research Center (CCRC), King Abdullah University of Science and Technology, Thuwal, Western Province, Saudi Arabia

Anqi Zhang
Aramco Americas: Aramco Research Center—Detroit, Novi, MI, United States

Yu Zhang
Aramco Americas: Aramco Research Center—Detroit, Novi, MI, United States

Le Zhao
Energy Systems Division, Argonne National Laboratory, Lemont, IL, United States

Foreword

The internal combustion engine (ICE) powers our world. From cars and long-haul trucks to agricultural and construction equipments, liquid- and gaseous-fueled engines touch practically every aspect of our lives. ICEs have been around for two hundred years, and the vehicles they power permeated the masses more than a century ago. These machines have improved our lives by lifting people out of poverty and giving us the freedom to live and work where we please.

Today's ICE looks vastly different from an 1800s-era engine. Significant advances in air delivery, fuel delivery, and emissions control systems have made the modern engine a much cleaner and more efficient machine. Even with these advances, however, the combustion engine is not sustainable in its current form. There is now a global race for decarbonization, which is leading to the development of new innovations at an unprecedented pace.

The ICE no longer has to go alone. Technologies such as battery electric and hydrogen fuel cell vehicles hold great promise for helping us achieve a cleaner, greener, and more diverse mobility future. Practically, though, these technologies can be only *part* of the solution. We still need engines—a lot of them—but the engine of tomorrow must be different from today's machines; the engine of tomorrow must be decarbonized. Achieving a fully decarbonized engine requires improvements in engine technologies, hybridization, and low- or no-carbon fuels.

How do we get there? The path forward will be a combination of improved simulation through techniques such as computational fluid dynamics and advances in engine hardware and controls. Virtualization through simulation has been, and continues to be, powerful for optimizing new engine technologies. As an example, the Mazda SKYACTIV-X was developed virtually before any physical prototypes were built, and this success is just the start of what is to come. But simulation methods alone are not enough to get us where we need to go, at least not in the short amount of time we have to get there.

The engine represents a large and complicated design space, encompassing fuels, combustion chamber design, fuel injection, ignition strategies, and more. This is a massive, multi-objective optimization problem with not just one, but a collection of optima. Making the most of this parameter space will require continued advances in engine control technologies, sensors, and onboard computers. These advances are already leading to unprecedented opportunity, but also to an ever-expanding and unmanageable design space for modern engines. Current trends are showing an exponential increase in the parameter space, and this trend is expected to continue for the foreseeable future. How can we possibly evaluate and implement every design combination to find the best solution? The current inability to realize the potential of this parameter space efficiently and effectively is leading to suboptimal engines in the market.

Fortunately, "big science" tools have evolved alongside the engine itself. Supercomputing, advanced numerical methods, and even artificial intelligence (AI) are now at our disposal. As with so many disciplines today, machine learning (ML) has opened a new realm of possibilities for advancing engines. By marrying big science/data science with virtual representations of the engine, we can (and will!) achieve the ultimate ICE design. AI/ML will bridge atomistic modeling breakthroughs with engineering simulations, and AI/ML controls will also enable autonomous, intelligent systems controls. These systems will have the ability to learn, adapt, and manipulate engine controls to operate at the edge of stability to maximize efficiency and minimize emissions under ever-changing vehicle demands.

This book does an excellent job covering the relatively new topic of AI applied to engine simulation and experimentation. Instead of trying to cover the entire design space at once, the authors of each chapter focus on a single aspect, such as fuel formulation, engine calibration parameters, combustion chamber design, detecting abnormal engine phenomena (e.g., low pressure ignition), and more. Furthermore, the authors do not focus solely on the current learning algorithms—they also provide a historical perspective and cover optimization methods based on the principles of evolution, for example, genetic algorithms.

This book provides a thorough and timely overview of where AI-assisted engine optimization stands today, making it a valuable reference for new researchers and seasoned engine designers alike. The editors have succeeded in choosing not only the right topics to cover but also the right experts to write about them.

We are entering a brave new world for the ICE. Our engines must be cleaner and greener than ever to stay relevant in today's decarbonizing society. This book lays out the most promising tools we have available to ensure the ICE reaches its full potential. Through the right combination of human intelligence and AI, advanced engines will help us decarbonize transportation.

Dr. Kelly Senecal
Co-founder, Convergent Science
Madison, WI, United States

Dr. Robert Wagner
Director, Buildings and Transportation Science Division,
Oak Ridge National Laboratory
Knoxville, TN, United States

Preface

Internal combustion engines are not going to go away any time soon. Internal combustion engine is currently the principal prime mover in the transportation sector, and it is expected to maintain its dominance in the foreseeable future. We therefore, strongly believe that improving the efficiency and reducing the emissions from internal combustion engines are the most effective measures to have an immediate and prominent impact on the environment. Our aim in this text is to provide an exposition of various engine and fuel optimization techniques and to showcase some of their applications. Specifically, the design optimization using advanced artificial intelligence and data-driven methods is the focus of this text. We hope that the readers will enjoy it and, more importantly, learn from it.

The text is primarily adapted from previous archival publications made by the various active research groups in this specific area. The authors list is a combination of leading, reputable, and active experts in the field of internal combustion engines and artificial intelligence. We are grateful to the authors for their invaluable contributions and are truly honored to have worked with such an elite group of world-renowned researchers.

Jihad Badra, Team Leader
Engine Combustion Team, Transport Technologies Division, Research and Development Center, Saudi Aramco, Dhahran, Eastern Province, Saudi Arabia

Pinaki Pal, Research Scientist
Energy Systems Division, Argonne National Laboratory, Lemont, IL, United States

Yuanjiang Pei, Team Leader
Computational Modeling Team, Aramco Americas: Aramco Research Center—Detroit, Novi, MI, Unites States

Sibendu Som, Manager
Multi-Physics Computation Section, Energy Systems Division, Argonne National Laboratory, Lemont, IL, United States

CHAPTER 1

Introduction

Balaji Mohan[1], Pinaki Pal[2], Jihad Badra[1], Yuanjiang Pei[3] and Sibendu Som[2]

[1]Transport Technologies Division, Research and Development Center, Saudi Aramco, Dhahran, Eastern Province, Saudi Arabia; [2]Energy Systems Division, Argonne National Laboratory, Lemont, IL, United States; [3]Aramco Americas: Aramco Research Center—Detroit, Novi, MI, United States

1. Industrial revolution

The first industrial revolution used steam power for production and mechanization. The second used electric power, which also led to mass production. The advancements in electronics and information technology led to the third industrial revolution. Now building upon the third, the digital revolution is taking place from the late last century. The technology breakthroughs in fields such as artificial intelligence (AI), big data analytics, internet of things, robotics, autonomous vehicles, unmanned aerial vehicles, three-dimensional (3D) printing, modeling, intelligent sensing, cloud computing, mobility, and augmented/virtual reality are multiplying the enormous possibilities to advance industries to make them smarter [1]. Fig. 1.1 shows the stages of revolution the industrial landscape has gone through over the period. AI technologies driven by big data will fuel the fourth industrial revolution. Although AI's roots can be traced back to the 1970s, its substantial impact on the latest industrial revolution has been evident over the last one to two decades.

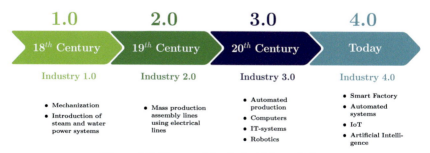

Figure 1.1 Stages of the industrial revolution.

2. Artificial intelligence, machine learning, and deep learning

AI has already occupied an essential place in our lives, from self-driving cars to virtual assistants. The exponential increase in computing power over the last decades and the vast amount of data collected through the start of the third revolution have paved the path for AI development and progress in recent years. The terms AI, machine learning (ML), and deep learning (DL) are mostly used interchangeably. However, these terms are different by definition. Fig. 1.2 shows the hierarchy of AI, ML, and DL. AI is the science of making things smart. In broad terms, it is a discipline to make computers perform complex human tasks, solve complex problems, and make intelligent decisions [2]. ML is a subset of AI, and it is an approach to achieve AI through systems that can learn from experience to find patterns in datasets. In 1959, Arthur Samuel, a computer scientist, coined the term "machine learning," which is defined as "computer's ability to learn without being explicitly programmed" [3]. The ML algorithms take known data, understand patterns, classify, cluster, or predict new data [4]. On the other hand, DL is a subset of ML, which generally uses neural networks with a structure similar to human neural systems to analyze and solve a particular problem [5].

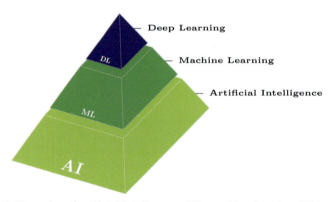

Figure 1.2 Hierarchy of artificial intelligence (AI), machine learning (ML), and deep learning (DL).

3. Machine learning algorithms

There are different ML algorithms based on their applications. The algorithms can be classified into three categories based on their learning styles—supervised, unsupervised, and reinforcement learning, as shown in Fig. 1.3 [6–8]. In supervised learning, the algorithms are trained using labeled data to generate a function that maps the targets to the inputs. The labeled data pair the inputs with their respective targets. The training process is continued until the model achieves a desired accuracy on the training data. Some commonly used supervised ML algorithms are linear regression, logistic regression, decision trees, Naïve Bayes, neural networks,

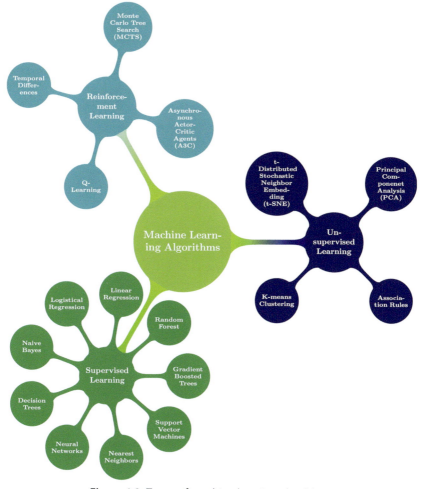

Figure 1.3 Types of machine learning algorithms.

nearest neighbors, support vector machines, gradient boosted trees, and random forest. In unsupervised learning, the algorithms use unlabeled data to cluster or segment them into different groups by identifying specific patterns. K-means clustering, principal component analysis, association rules, and t-distributed stochastic neighbor embedding are commonly used unsupervised learning algorithms. The reinforcement learning algorithm iteratively learns from its environment. In the process, the algorithm also referred to as an agent, learns from its experience until it explores the full range of possible states. It works on a simple feedback system to maximize the reward or minimize the risk. This feedback is called a reinforcement signal. Some commonly used reinforcement learning algorithms are Q-learning, temporal differences, Monte Carlo tree search, and asynchronous actor-critic agents.

There are many applications for ML algorithms in day-to-day lives, for example, image recognition, virtual assistants like SIRI, Alexa, and Cortana, speech predictions for search by voice technology, traffic prediction based on real-time location data, product recommendations by various e-commerce websites, self-driving cars, email spam, and malware filtering, stock market diagnosis, automatic language translation, and medical diagnosis. The applications of ML algorithms have further expended to optimize the performance and emissions from internal combustion engines (ICEs).

4. Artificial intelligence—based fuel-engine co-optimization

AI-based co-optimization of ICE and fuel includes optimizing engine design and operating parameters, fuel formulation, and properties, and mitigating the unwanted and potentially damaging combustion events. Optimization of fuel formulation includes high-throughput screening, prediction of fuel and mixture properties, fuel formulation, and developing reaction mechanisms for fuels. In addition, AI-based approaches can predict abnormal combustion events such as preignition to assist manufacturers in developing engines susceptible to this phenomenon and diagnose the preignition during on-board diagnosis and mitigate it. The following sections will review the various applications in detail.

4.1 Optimization of internal combustion engine

ICEs have been around since the 19th century, and their conceptual identity as a fuel-powered machine has not changed since. Despite the complexity of ICEs, there have been significant technological improvements to their performance in response to fuel efficiency and emission regulations [9]. Many parameters affect the performance and emission characteristics of an ICE. These include but not limited to, fuel characteristics [10,11], exhaust gas recirculation (EGR) [12−14], piston bowl geometry [15,16], injection strategy [17−23], injection timing [19,24,25], spray angle [26−28], and swirl ratio [20,29,30]. Although in-depth studies of a single parameter may provide valuable insights into its effects on engine performance, the interactions between the various input parameters and output variables are interlinked, nonlinear, and complex.

The tools used to co-optimize the fuel/engine system have evolved over the years. Until 20 years ago, experimental prototyping (manual) was the main optimizing method. It was followed by numerical simulations, including complex 3D computational fluid dynamics (CFD), which played a major role in engine/fuel system optimization. This development was enabled by the significant advancements in computing power (supercomputers, clusters, and parallelization) and numerical models (turbulence, combustion, spray, heat transfer, meshing, and moving boundaries). Due to high dimensionality, complexity, and highly nonlinear dependencies in engine properties and responses, both experimental and numerical optimization approaches can be inefficient and take a significant amount of time and effort to obtain local rather than global optimum designs and operating conditions [31−33]. Alternative approaches have been developed over the years to overcome the issues with manual optimizations. These methods in chronological order include the design of experiments (DoE), genetic algorithms (GAs), ML, and AI, and the evolution along with the high-level enablers and limitations are shown in Fig. 1.4. More details about each of the recent optimization techniques are provided in the following sections.

4.1.1 Design of experiments

In the early days, engine optimization was performed manually using very few designs, mainly due to limited computational power and resources. Later, the DoE-response surface methodology (RSM) optimization method became very popular and was used widely in engine optimization due to advancements in the field of information technology [34−39]. It uses statistical techniques to build proper response surfaces for optimization. The DoE-RSM optimization is considered the industry-standard practice as they provide a quick solution with fewer resources. For instance, Besson et al. [40] from Renault used the CFD-DoE strategy to optimize the homogenous charge compression ignition (HCCI) combustion chamber under full load operating point using a DoE size of 30. Han et al. [41] from Ford used two-piston designs and two injector configurations to optimize the intake

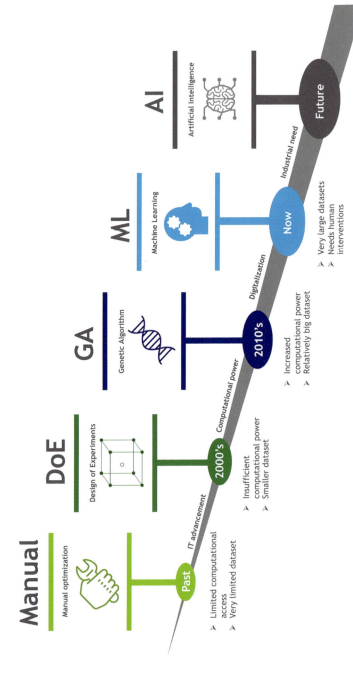

Figure 1.4 Evolution of engine performance and emission optimization strategies.

and spray injection patterns using the CFD-DoE optimization approach of a new light stratified charge direct injection spark ignition (DISI) combustion system. Lippert et al. [42] from General Motors and Suzuki Motor Corporations jointly optimized piston and injector design using CFD-DoE comprising three-piston designs and seven injector configurations of a small displacement spark-ignited (SI) direct injection (DI) engine under stratified operation. Hajireza et al. [37] from AVL used a DoE size of 16 designs to optimize the piston bowl, spray cone angle, and intake port swirl level of a diesel engine using CFD. Davis et al. [43] from General motors and Robert Bosch jointly developed the 3.6L DOHC 4V V6 DI engine's combustion system using the CFD-DoE optimization approach from six-piston designs and three injector configurations. Reiche et al. [44] from Ford experimentally optimized the cold start for the EcoBoost engine using the DoE technique with three pistons and four-port designs. Catania et al. [45] used experimental-DoE to optimize new combustion system development specifically oriented toward premixed charge compression ignition with three-piston designs. Styron et al. [38] used the CFD-DoE approach to optimize the Ford 2011 6.7L Power Stroke diesel engine with a DoE size of 16 designs comprising three pistons. Rajamani et al. [36] performed a parametric analysis of piston bowl geometry and injection nozzle configuration using CFD-DoE with 16 designs. Pei et al. [39] from Aramco Research Center collaborated with Argonne National Lab and Convergent Science Inc. to optimize the piston bowl geometry, the number of injector nozzles, total nozzle area, nozzle inclusion angle, and the start of injection (SOI) of a heavy-duty (HD) engine using a comprehensive CFD-DoE approach. Later, Pei et al. [15] used CFD-DoE comprising 256 combustion chamber designs to optimize injector spray patterns, fuel injection strategies, and in-cylinder swirl motion to achieve better fuel efficiency from a gasoline compression ignition (GCI) engine. Probst et al. [34] used the CFD-DoE-GA approach to optimize fuel consumption and NO_x emissions from a diesel engine.

4.1.2 Genetic algorithm
The advancements in computational power enabled GA [46–48] to evolve as a popular optimization approach among researchers. GA typically has a better chance of finding global optimum solutions despite the presence of multiple local minima. It is widely used in academia due to the considerable resources and time required to develop an efficient approach to better solve a problem. For instance, Senecal and Reitz [49] optimized a DI diesel

engine's emissions and fuel consumption simultaneously using the CFD-GA approach. Risi et al. [50] used CFD-GA to select a combustion chamber, giving the best compromise of the selected fitness factors based on the engine emissions levels and a penalty function to account for engine performance. Shi and Reitz [51] used CFD-GA to study the effects of bowl geometry, spray targeting, and swirl ratio for a HD diesel engine operating at a high load. The simultaneous reduction in emissions and improved fuel consumption was achieved in this optimization study. Ge et al. [52] optimized the high-speed DI diesel engine's SOI, swirl ratio, boost pressure, and injection pressure using the CFD-GA approach to reduce fuel consumption and pollutant emissions simultaneously. Hanson et al. [53] optimized the reactivity controlled compression ignition combustion in a light-duty multicylinder engine under three operating conditions in the US Environmental Protection Agency light-duty Federal Test Procedure (FTP) test to improve thermal efficiency while maintaining low nitrogen oxide (NO) emissions. Bertram et al. [51] used experiments combined with GA and particle swarm optimization (PSO) to optimize a single objective function representing oxides of nitrogen (NO_x), particulate matter (PM), hydrocarbon (HC), carbon monoxide (CO), and fuel consumption of a diesel engine. Zhang et al. [52] combined experiments with GA and PSO to optimize a diesel engine fueled with soy biodiesel. The dynamometer time, fuel economy, and exhaust emissions were improved using the hybrid GA-PSO algorithm. Recently, Broatch et al. [54] combined CFD modeling and GA technique to optimize the combustion system hardware design of a high-speed DI diesel engine, with respect to various emissions and performance targets, including combustion noise.

4.1.3 Machine learning–based algorithms

With advancements in digitization and AI algorithms, ML has emerged as an effective tool for optimizing complex systems, such as engine combustion concepts. Many studies have shown that ML-based optimization approaches perform better than the RSM-based method when the dataset's complexity increases [55,56]. Further, the ML models can be used as surrogates to represent/replace the engine system. Artificial neural networks (ANNs) have been utilized to optimize engine performances and emissions or represent the engine system as a surrogate model. For instance, Traver et al. [57] employed neural networks to predict NO_x and carbon dioxide (CO_2) emissions using in-cylinder pressure-based variables. Hafner et al. [58] used fast neural network models for diesel engine control design.

The neuromodels were integrated with an upper-level emission optimization tool to determine the optimal engine settings to minimize the emissions. Desantes et al. [59] used HD diesel engine operating conditions as design parameters to neural networks to predict exhaust emissions, such as NO_x and PM. Furthermore, they optimized the engine to meet EURO IV emissions standards. Brahma and Rutland [60] used neural networks to predict the pressure, temperature, heat flux, torque, and emissions from a diesel engine's operating parameters. They integrated neural network models with a GA combined with a hill-climbing strategy to optimize the engine's entire speed-torque map's operating parameters. The neural networks acted as an engine map to simulate the engine's emissions over the FTP HD diesel cycle with reasonable accuracy and short computational time. He and Rutland [61] used ANNs to optimize diesel engine emissions using CFD. Seven control parameters, namely, engine speed, engine load, the SOI, injection pressure, fuel mass in the first injection, boost pressure, and EGR, were used as inputs to the neural networks. The neural networks predicted five objective parameters: cylinder pressure, cylinder temperature, cylinder wall heat transfer, NO_x, and soot emissions. Kesgin [9] used a neural network and a simple GA to predict and optimize the effects of design and operational parameters on natural gas engine efficiency and NO_x emissions. Wu et al. [62] used ANN as a surrogate model representing the engine's response to different control variable combinations of variable valve actuation technology with greatly reduced computational time. The optimal cam-phasing strategy obtained at wide-open throttle for a dual independent variable valve timing (VVT) engine was well-validated with engine dynamometer tests. Anand et al. [63] used ANN to predict SI engine efficiency and NO_x emission based on engine design and operational parameters trained on quasi-dimensional, two-zone thermodynamic simulation data. The ANN model was suggested as an alternative to the model-based simulations for real-time computations in the electronic control unit (ECU) due to its good accuracy and low computational time. Atashkari et al. [64] used SI engine experimental data comprising intake valve timing and engine speed as inputs and engine torque and fuel consumption as outputs to train group method of data handling type neural network. Then, a nondominated sorting genetic algorithm (NSGA-II) was used to optimize the VVT SI engine's performance. Ashhab [65] used feedforward ANN to predict the engine outputs, pumping loss, and cylinder air charge using intake valve lift and closing time as inputs. This neural network model developed a camless engine inverse control algorithm, which performed better than the graphical-based techniques. Cruz-Peragon [66] used a combination of angular speed measurements and ANN for combustion fault diagnosis in ICE. The experimental data from three different combustion engines were utilized for training the ANN-based combustion diagnosis tool to predict defects in pressure waveforms and other combustion indicators, such as fuel consumption and ignition timing. Kiani et al. [67] used ANN to predict the engine brake power, torque, and exhaust emissions, such as CO, CO_2, NO_x, and HC. The ANN was trained using experimental data from SI engine fueled by ethanol-gasoline blends in different percentages (0%, 5%, 15%, and 20%) operated at different engine speeds and loads. The authors suggested that the ANN could be used as a surrogate model due to its advantage of fast, accurate, and reliable predictions when other numerical and mathematical models fail. Tasdemir et al. [68] modeled the SI engine performance and emissions using ANN and fuzzy expert system. The model was applied to predict the engine power, torque, specific fuel consumption (SFC), and HC emissions. The model performed well in predicting the engine performance and emissions of new experiments. Barma et al. [69] predicted the diesel engine parameters such as mean effective pressure, efficiency, fuel consumption, air-fuel ratio, and torque from engine speed, load, and different biodiesel and diesel fuel blends as inputs using a back propagation-based ANN. The model showed better accuracy in predicting engine performance fueled by biodiesel and diesel fuels. Yap and Karri [70] developed a two-stage ANN where the first stage predictions of power and tractive forces were used as the second stage ANN inputs to predict CO, CO_2, HC, and oxygen (O_2) from a scooter engine. The model was suggested to be used as a virtual emission sensor without additional equipment and further substantiates that only a single ECU was required to predict engine performance and emissions. Çay et al. [71] used ANN to predict the brake-specific fuel consumption (BSFC), air-fuel ratio, CO and HC emissions from engine speed, torque, fuel flow rate, and fuel blend ratios of a methanol-gasoline blend fueled gasoline engine. The ANN model's predictions showed that methanol as a fuel resulted in lower emissions while gasoline fuel led to better engine performance, and the experimental results validated the predictions. Taghavifar et al. [72] trained an ANN model to act as a surrogate model to predict spray characteristics, such as liquid penetration length and Sauter mean diameter. The ANN model was trained with a combination of experimental and CFD simulation data of spray characteristics where engine crank angle, vapor mass flow rate, turbulence, and nozzle outlet pressure were used as inputs to the model. GA was used to optimize the network's interconnecting weights to achieve high accuracy. Rezaei et al. [73] developed radial basis function (RBF)-based ANN as a simulation testbed for performance analysis, calibration, and testing of HCCI engine controllers fueled by ethanol and butanol. The model predicted the indicated mean effective pressure, indicated thermal efficiency, total net heat released, maximum in-cylinder pressure, HC, NO_x, and CO from butanol volume percentage, and fuel equivalence ratio as inputs. The RBF-based ANN performed well compared to the feed-forward ANN in terms of prediction accuracy. Bietresato et al. [74] predicted the instant torque and BSFC of a farm tractor diesel engine with only exhaust gas temperature as an input. It was further found that the relation of lubricant temperature with instant torque and BSFC was poor and resulted in a very low and diversified coefficient of determination (R^2). ANN proved to be a useful and reliable tool for correlating exhaust temperature to instant torque and BSFC. Therefore, it was suggested that it could be effectively used as an engine monitoring system. Kapusuz et al. [75] demonstrated an ANN to be successfully used as an alternative type of modeling technique for SI engines powered by alcohol-unleaded gasoline mixtures. The model was trained with experimental data comprising engine speed and methanol/ethanol percentages to predict torque, power, and BSFC. Shamekhi and Shamekhi [76] developed a real-time control-oriented model with high precision for a multipoint fuel injection SI engine using neural networks. They termed it the neuro-mean value model (Neuro-MVM). The Neuro-MVM acted as a gray-box model and performed better than both plain MVM (white-box model) and mere neural networks (black-box model) regarding accuracy and reliability in predictions. Rahimi-Groji et al. [77] developed an ANN and regression-based model to predict power and indicated specific fuel consumption, respectively, to tune up ECU under different climatic conditions instead of running the engine dynamometer to save cost and time. The models were trained based on the engine's one-dimensional model, which was well-validated using experimental data. Bahri et al. [78] demonstrated that the combustion noise level (CNL) in HCCI engines could be modeled with an ANN. The developed ANN noise level (ANL) model predicted the CNL in real-time with less than 0.5% error for the HCCI engine. The ANL model could identify the engine operating limit to avoid ringing operation. The literature shows that the ANN models have been widely used to predict various performance, emissions, and other ICE parameters. Although the advantage of ANNs over other ML approaches is the ability to work with inadequate information once the models have been well trained, they require greater computational resources, are prone to overfitting, and lose effectiveness when extrapolating beyond the original data [79,80].

Recently, more sophisticated ML algorithms have been developed and demonstrated for engine optimization. Togun and Baysec [81] used decision trees combined with genetic programming (GP) to predict torque and BSFC of a gasoline engine in spark advance, throttle position, and engine speed. Compared to an ANN model, the decision trees combined with GP were more accurate, faster, and more practical. Moiz et al. [48] used SuperLearner, an ensemble ML algorithm combined with a GA, to optimize a HD GCI engine. Later, the SuperLearner approach combined with a grid gradient ascent method was employed to optimize the piston bowl geometries of two GCI engines [82,83]. These studies demonstrated the potential of an accelerated CFD-based ML optimization approach to improve engine performance. ML-GA enables the user to better predict by optimizing parameters such as training dataset size, fitting models, and fitting models' hyperparameters. More recently, Owoyele et al. [84,85] introduced an automated ML-GA technique (AutoML-GA), which improves ML-GA by incorporating automated hyperparameter tuning with an active learning loop. In this framework, the process of selecting hyperparameters manually was replaced by a Bayesian approach. It was shown that the hyperparameters chosen in this fashion led to more accurate surrogate models and a shorter active learning loop. On the other hand, Owoyele and Pal [86,87] demonstrated a novel ensemble ML-driven Active Optimizer (ActivO) technique to significantly accelerate engine design optimization. In this approach, two different ML surrogates, called *weak* and *strong* learners, trained in an active learning loop, performed exploration and exploitation of the large design space. The *weak* learner successively guided the optimizer toward the promising regions within the design space, and the *strong* learner was used to exploit those promising regions to find the global optimum.

Fig. 1.5 shows a summary of the techniques employed for the design optimization of ICEs. The waffle chart on the left shows the design parameters usually incorporated in the optimization campaign. Each box's size indicates the number of times the parameter has been included as a design parameter in optimization. The middle pie chart shows different optimization algorithms used and their relative predominance. The parameters usually considered as the optimization targets are shown in the

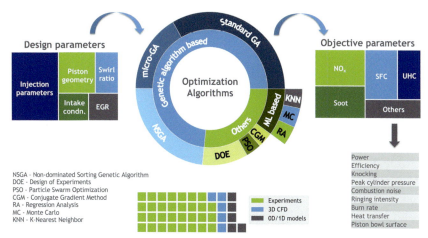

Figure 1.5 Summary of the machine learning (ML)-based optimization of internal combustion engines.

right waffle chart. The objectives were generally performance parameters, like SFC and emissions (NO$_x$, Soot, and unburned hydrocarbon). The bottom waffle chart shows the proportion of experiments, 3D-CFD simulations, and 0/1D simulations to which optimization algorithms have been applied to optimize the engine's performance and emissions.

4.2 Optimization of fuel formulation

Liquid HC fuels are complex mixtures of hundreds of different molecules. The specifications and properties of commercial liquid HC fuels such as market gasoline and diesel are set to meet emission regulations and optimize refinery economics. It is widely accepted that the fuel properties significantly affect engine performance and emissions [19,88–95]. Therefore, the co-optimization of the engine and fuel is necessary to unlock the full potential of ICEs. Like the engine parameters, the relationship between fuel properties and engine performance and emissions is very complex, and AI can be an excellent enabler to optimize the fuel properties. Various groups have already attempted to optimize transport fuels using advanced AI methods. High throughput screening, prediction of fuel properties, fuel formulation, and developing reaction mechanisms are some of the AI-enabled fuel optimization applications.

DL approaches combined with gradient-based optimization are used to generate molecules from the most probable latent space decoding [96,97].

Recurring neural networks (RNNs) can generate simplified molecular-input line-entry system (SMILES) through a decoder, and generative adversarial networks (GANs) can directly generate molecular graphs by combining a generator and a discriminator network. RNN-based reinforcement learning algorithm allows performing genuine quantitative structure-property relationships analysis in which one part of the network generating SMILES and the other component is used to predict its properties [98]. Unlike RNN, GAN explicitly exempts users from knowing probability distribution, which is interesting for high-dimensional data [99].

Usually, the fuel mixture properties are deduced through the mixing rule at the molecular descriptor or property level from the pure species. AI helps in predicting fuel properties of the pure species through some molecular descriptors common to multiple properties. However, in the absence of detailed composition, fuel properties may be correlated to readily available observables [100]. For example, American Society for Testing Materials (ASTM) distillation curves [101], Fourier transforms infrared [102], flame emission [103], nuclear magnetic resonance (NMR) [104], dispersive fiber-optic Raman [105], and dielectric spectroscopy [106] were processed to generate discriminative input features by different groups. Then, the spectra are either directly used or refined to construct the functional groups, which are then used as input features to predict properties. Interestingly, while ANN modeling of RON and MON of gasoline using nine structural descriptors (paraffinic primary to tertiary carbons, olefinic, naphthenic, aromatic and ethanolic OH groups, molecular weight, and branching index) obtained from ^{1}H NMR spectra was successful (MAE = 1.2, R^2 = 0.99), multiple linear regression training over the same dataset with same input features resulted in a poor correlation with R^2 = 0.51 [107]. This comparison demonstrated the nonlinearity of RON and MON with respect to these structural descriptors. ANN can be effectively used for octane prediction that leverages both detailed chemical kinetics for ignition calculations and chemical structure information to achieve high accuracy predictions of RON and MON. Then a fuel search task can be accomplished as an optimization problem to maximize the load range of mixed-mode operation of the fuel composition, which also satisfies constraints on RON. Further details of this fuel search optimization can be found in Chapter 3.

Formulating reaction mechanisms is an essential step in fuel formulation, which provides adequate combustion thermokinetic models for numerical engine design using CFD and refinement of fuel property accuracy.

AI helps in the search of optimal reaction pathways using reinforcement learning techniques. ANN has been used to achieve chemical accuracy, i.e., error on electronic energy <1 kcal/mol [108]. ANN has also shown to be much more accurate than their conventional counterparts in predicting reaction force fields [109]. In any case, AI-based methods are not expected to supersede physical methods that seek to explicitly understand the inner workings of a phenomenon. In contrast, data-driven strategies focus on approximating its outer behavior [110].

4.3 Mitigation of rare combustion events

AI can also be used to predict and prevent abnormal combustion events that limit engine performance. For example, low-speed preignition (LSPI), also known as stochastic preignition, is an abnormal, potentially destructive combustion phenomenon in gasoline SI engines when the air-fuel mixture is ignited prematurely. LSPI most commonly occurs in turbocharged direct-injection engines that are operating at low speeds and high loads. Many manufacturers are developing engines susceptible to this phenomenon to meet ever-increasing and challenging fuel efficiency and emission regulations.

ML and DL models have been proposed for vehicle fault diagnostics. These include misfire detection in conventional ICEs [111] and dynamic skip-fired engines [112]. Gunnemann and Pfeffer [113] addressed the problem of engine damage prediction using deep convolutional neural networks from structure-borne noise. Yuan et al. [114] employed RNNs for fault detection and estimation of the remaining useful life of an aero engine. Zhu et al. [115] used ANN to predict the preignition phenomenon using the ion current/cylinder pressure cooperative combustion data obtained from 154,077 valid cycles. The results confirmed that the ion current signal could give additional information beyond the pressure trace. Kodavasal et al. [116] used the random forest to learn the relationship between metrics representing flame topology and preignition flow-fields and peak cylinder pressure using a dataset comprised of multiple cycles of large-eddy simulation calculations performed on a spark ignited gasoline engine with a cycle-to-cycle variation (CCV). Kuzhagaliyeva et al. [117] developed a DL approach to use data from a lambda sensor and a low-resolution exhaust back pressure sensor to detect preignition events for eventual deployment in on-road preignition monitoring. Schuman et al. [118] demonstrated a complete neuromorphic workflow, from application to hardware, for training and deploying low size, weight, and power neuromorphic solutions to control and improve fuel efficiency in spark-ignition ICEs. A hybrid strategy combining AI components within a physics-informed framework referred to as "gray-box" can be used effectively to predict and control the abnormal combustion phenomena. This hybrid strategy can be used to control CCV in dilute spark-ignition engines, combustion stability control, and identify dilute limit changes in HCCI engines based on a spark/EGR sweep to avoid misfires. Further details about this gray-box strategy can be found in Chapter 8.

5. Summary

This chapter has reviewed the application of AI in the co-optimization of ICEs and fuels, including optimizing engine design and operating parameters, optimizing fuel formulation, and mitigating rare combustion events. AI-based techniques enable engine manufacturers to develop new-generation engines and fuels with high efficiency, ultralow emissions, and susceptibility to rare combustion events such as preignition and knocking. The clear advantages of ML-based optimization over traditional approaches make it an attractive alternative to achieve a quick and reliable solution. However, there is still human participation required in various stages of the optimization process. Smart industries need to automate further and improve the entire process's overall efficiency. By fully integrating the process through multiple elements of industrial revolution 4.0, the smart industry's real-world challenges can be solved more quickly and efficiently using less resources.

References

[1] Schwab K. The fourth industrial revolution. Worl Economic Forum; 2017.
[2] Nilsson NJ. Principles of artificial intelligence. Morgan Kaufmann; 2014.
[3] Samuel AL. Some studies in machine learning using the game of checkers. IBM J Res Dev 1959;3(3):210—29.
[4] Langley P. Elements of machine learning. Morgan Kaufmann; 1996.
[5] LeCun Y, Bengio Y, Hinton G. Deep learning. Nature 2015;521(7553):436—44.
[6] Bonaccorso G. Machine learning algorithms. Packt Publishing Ltd; 2017.
[7] Ayodele TO. Types of machine learning algorithms. N Adv Machine Learn 2010;3:19—48.
[8] Mohammed M, Khan MB, Bashier EBM. Machine learning: algorithms and applications. Crc Press; 2016.
[9] Kesgin U. Genetic algorithm and artificial neural network for engine optimisation of efficiency and NOx emission. Fuel 2004;83(7—8):885—95.

[10] Vallinayagam R, An Y, Vedharaj S, Sim J, Chang J, Johansson B. Naphtha vs. dieseline—The effect of fuel properties on combustion homogeneity in transition from CI combustion towards HCCI. Fuel 2018;224:451—60.
[11] Vallinayagam R, Hlaing P, AlRamadan AS, An Y, Sim J, Chang J, Johansson B. The physical and chemical effects of fuel on gasoline compression ignition. SAE Technical Paper; 2019.
[12] Jiang C, Huang G, Liu G, Qian Y, Lu X. Optimizing gasoline compression ignition engine performance and emissions: combined effects of exhaust gas recirculation and fuel octane number. Appl Therm Eng 2019;153:669—77.
[13] Pan M, Qian W, Wei H, Feng D, Pan J. Effects on performance and emissions of gasoline compression ignition engine over a wide range of internal exhaust gas recirculation rates under lean conditions. Fuel 2020;265:116881.
[14] Zhou L, Hua J, Liu F, Liu F, Feng D, Wei H. Effect of internal exhaust gas recirculation on the combustion characteristics of gasoline compression ignition engine under low to idle conditions. Energy 2018;164:306—15.
[15] Pei Y, Pal P, Zhang Y, Traver M, Cleary D, Futterer C, Brenner M, Probst D, Som S. CFD-guided combustion system optimization of a gasoline range fuel in a heavy-duty compression ignition engine using automatic piston geometry generation and a supercomputer. SAE Int J Adv Curr Prac Mobility 2019;1(1):166—79. https://doi.org/10.4271/2019-01-0001.
[16] Tang M, Pei Y, Guo H, Zhang Y, Torelli R, Probst D, Futterer C, Traver M. Piston bowl geometry effects on gasoline compression ignition in a heavy-duty diesel engine. 2020.
[17] Benajes J, Molina S, García A, Monsalve-Serrano J, Durrett R. Performance and engine-out emissions evaluation of the double injection strategy applied to the gasoline partially premixed compression ignition spark assisted combustion concept. Appl Energy 2014;134:90—101.
[18] Kim D, Bae C. Application of double-injection strategy on gasoline compression ignition engine under low load condition. Fuel 2017;203:792—801.
[19] Atef N, Badra J, Jaasim M, Im HG, Sarathy SM. Numerical investigation of injector geometry effects on fuel stratification in a GCI engine. Fuel 2018;214:580—9.
[20] Kodavasal J, Kolodziej CP, Ciatti SA, Som S. Effects of injection parameters, boost, and swirl ratio on gasoline compression ignition operation at idle and low-load conditions. Int J Engine Res 2017;18(8):824—36.
[21] Mohan B, Yang W, Chou Sk. Fuel injection strategies for performance improvement and emissions reduction in compression ignition engines—a review. Renew Sustain Energy Rev 2013;28:664—76.
[22] Mohan B, Yang W, Yu W, Tay KL, Chou SK. Numerical investigation on the effects of injection rate shaping on combustion and emission characteristics of biodiesel fueled CI engine. Appl Energy 2015;160:737—45.
[23] Tay KL, Yang W, Zhao F, Yu W, Mohan B. Effects of triangular and ramp injection rate-shapes on the performance and emissions of a kerosene-diesel fueled direct injection compression ignition engine: a numerical study. Appl Therm Eng 2017;110:1401—10.
[24] Naser N, Jaasim M, Atef N, Chung SH, Im HG, Sarathy SM. On the effects of fuel properties and injection timing in partially premixed compression ignition of low octane fuels. Fuel 2017;207:373—88.
[25] Jaasim M, Hernandez Perez F, Vallinayagam R, Vedharaj S, Johansson B, Im HG. Computational study of stratified combustion in an optical diesel engine. SAE Technical Paper; 2017.

[26] Tang Q, An Y, Raman V, Shi H, Sim J, Chang J, Magnotti G, Johansson B. Experimental study on the effects of spray—wall interaction on partially premixed combustion and engine emissions. Energy Fuels 2019;33(6):5673—81.
[27] Tang Q, Liu H, Li M, Yao M. Optical study of spray-wall impingement impact on early-injection gasoline partially premixed combustion at low engine load. Appl Energy 2017;185:708—19.
[28] Mohan B, Ali MJM, Ahmed A, Perez FH, Sim J, Roberts W, Sarathy M, Im H. Numerical simulations of high reactivity gasoline fuel sprays under vaporizing and reactive conditions. SAE Technical Paper; 2018.
[29] Loeper P, Ra Y, Foster DE, Ghandhi J. Experimental and computational assessment of inlet swirl effects on a gasoline compression ignition (GCI) light-duty diesel engine. SAE Technical Paper; 2014.
[30] Zhang Y, Pei Y, Engineer N, Cho K, Cleary D. CFD-guided combustion strategy development for a higher reactivity gasoline in a light-duty gasoline compression ignition engine. SAE Technical Paper; 2017.
[31] Gen M, Cheng R. Genetic algorithms and engineering optimization. John Wiley & Sons; 1999.
[32] Manolas DA, Frangopoulos CA, Gialamas TP, Tsahalis DT. Operation optimization of an industrial cogeneration system by a genetic algorithm. Energy Convers Manag 1997;38(15—17):1625—36.
[33] Wong KI, Wong PK, Cheung CS, Vong CM. Modeling and optimization of bio-diesel engine performance using advanced machine learning methods. Energy 2013;55:519—28.
[34] Probst DM, Senecal PK, Chien PZ, Xu MX, Leyde BP. Optimization and uncertainty analysis of a diesel engine operating point using computational fluid dynamics. J Eng Gas Turbines Power 2018;140(10):102806.
[35] d'Ambrosio S, Ferrari A. Potential of double pilot injection strategies optimized with the design of experiments procedure to improve diesel engine emissions and performance. Appl Energy 2015;155:918—32.
[36] Rajamani VK, Schoenfeld S, Dhongde A. Parametric analysis of piston bowl geometry and injection nozzle configuration using 3D CFD and DoE. SAE Technical Paper; 2012.
[37] Hajireza S, Regner G, Christie A, Egert M, Mittermaier H. Application of CFD modeling in combustion bowl assessment of diesel engines using DoE methodology. SAE Technical Paper; 2006.
[38] Styron J, Baldwin B, Fulton B, Ives D, Ramanathan S. Ford 2011 6.7 L power stroke® diesel engine combustion system development. SAE Technical Paper; 2011.
[39] Pei Y, Zhang Y, Kumar P, Traver M, Cleary D, Ameen M, Som S, Probst D, Burton T, Pomraning E, et al. CFD-guided heavy duty mixing-controlled combustion system optimization with a gasoline-like fuel. SAE Int J Commer Veh 2017;10(2):532—46.
[40] Besson M, Hilaire N, Lahjaily H, Gastaldi P. Diesel combustion study at full load using CFD and design of experiments. SAE Technical Paper; 2003.
[41] Han Z, Weaver C, Wooldridge S, Alger T, Hilditch J, McGee J, Westrate B, Xu Z, Yi J, Chen X, et al. Development of a new light stratified-charge DISI combustion system for a family of engines with upfront CFD coupling with thermal and optical engine experiments. SAE Technical Paper; 2004.
[42] Lippert AM, El Tahry SH, Huebler MS, Parrish SE, Inoue H, Noyori T, Nakama K, Abe T. Development and optimization of a small-displacement spark-ignition direct-injection engine — stratified operation. SAE Technical Paper; 2004.

[43] Davis RS, Mandrusiak GD, Landenfeld T. Development of the combustion system for general motors' 3.6L DOHC 4V V6 engine with direct injection. SAE Int J Engines 2009;85−100.
[44] Reiche DB, Wooldridge ST, Moilanen PC, Davis GC. Experimental optimization of the cold start for the EcoBoost engine. SAE Technical Paper; 2009.
[45] Catania AE, d'Ambrosio S, Finesso R, Spessa E, Cipolla G, Vassallo A. Combustion system optimization of a low compression-ratio PCCI diesel engine for light-duty application. SAE Int J Engines 2009;2(1):1314−26.
[46] Awad OI, Mamat R, Ali OM, Azmi W, Kadirgama K, Yusri I, Leman A, Yusaf T. Response surface methodology (RSM) based multi-objective optimization of fusel oil-gasoline blends at different water content in SI engine. Energy Convers Manag 2017;150:222−41.
[47] Singh Y, Sharma A, Tiwari S, Singla A. Optimization of diesel engine performance and emission parameters employing cassia tora methyl esters-response surface methodology approach. Energy 2019;168:909−18.
[48] Moiz AA, Pal P, Probst D, Pei Y, Zhang Y, Som S, Kodavasal J. A machine learning-genetic algorithm (ML-GA) approach for rapid optimization using high-performance computing. SAE Int J Commer Veh 2018;11(5):291−306. https://doi.org/10.4271/2018-01-0190.
[49] Senecal PK, Reitz RD. Simultaneous reduction of engine emissions and fuel consumption using genetic algorithms and multi-dimensional spray and combustion modeling. SAE Technical Paper; 2000.
[50] de Risi A, Donateo T, Laforgia D. Optimization of the combustion chamber of direct injection diesel engines. SAE Technical Paper; 2003.
[51] Shi Y, Reitz RD. Assessment of optimization methodologies to study the effects of bowl geometry, spray targeting and swirl ratio for a heavy-duty diesel engine operated at high-load. SAE Int J Engines 2008;1(1):537−57.
[52] Ge H-W, Shi Y, Reitz RD, Wickman D, Willems W. Engine development using multi-dimensional CFD and computer optimization. SAE Technical Paper; 2010.
[53] Hanson R, Curran S, Wagner R, Kokjohn S, Splitter D, Reitz RD. Piston bowl optimization for RCCI combustion in a light-duty multi-cylinder engine. SAE Int J Engines 2012;5(2):286−99.
[54] Broatch A, Novella R, Gomez-Soriano J, Pal P, Som S. Numerical methodology for optimization of compression-ignited engines considering combustion noise control. SAE Int J Engines 2018;11(6):625−42.
[55] Maran JP, Priya B. Comparison of response surface methodology and artificial neural network approach towards efficient ultrasound-assisted biodiesel production from muskmelon oil. Ultrason Sonochem 2015;23:192−200.
[56] Oh S. Comparison of a response surface method and artificial neural network in predicting the aerodynamic performance of a wind turbine airfoil and its optimization. Appl Sci 2020;10(18):6277.
[57] Traver ML, Atkinson RJ, Atkinson CM. Neural network-based diesel engine emissions prediction using in-cylinder combustion pressure. J Fuel Lubric 1999;108(4):1166−80.
[58] Hafner M, Schüler M, Nelles O, Isermann R. Fast neural networks for diesel engine control design. Contr Eng Pract 2000;8(11):1211−21.
[59] Desantes JM, López JJ, García JM, Hernández L. Application of neural networks for prediction and optimization of exhaust emissions in a H.D. Diesel engine111. SAE Technical Paper; 2002. p. 1993−2002.
[60] Brahma I, Rutland C. Optimization of diesel engine operating parameters using neural networks. SAE Technical Paper; 2003. p. 2521−9.

[61] He Y, Rutland C. Application of artificial neural networks in engine modelling. Int J Engine Res 2004;5(4):281−96.
[62] Wu B, Prucka RG, Filipi ZS, Kramer DM, Ohl GL. Cam-phasing optimization using artificial neural networks as surrogate models—maximizing torque output. SAE Trans 2005;114:1586−99.
[63] Anand G, Gopinath S, Ravi MR, Kar IN, Subrahmanyam JP. Artificial neural networks for prediction of efficiency and NOx emission of a spark ignition engine. SAE Technical Paper; 2006.
[64] Atashkari K, Nariman-Zadeh N, Gölcü M, Khalkhali A, Jamali A. Modelling and multi-objective optimization of a variable valve-timing spark-ignition engine using polynomial neural networks and evolutionary algorithms. Energy Convers Manag 2007;48(3):1029−41.
[65] Ashhab MdSS. Fuel economy and torque tracking in camless engines through optimization of neural networks. Energy Convers Manag 2008;49(2):365−72.
[66] Cruz-Peragon F, Jimenez-Espadafor FJ, Palomar JM, Dorado MP. Combustion faults diagnosis in internal combustion engines using angular speed measurements and artificial neural networks. Energy Fuels 2008;22(5):2972−80.
[67] Kiani MKD, Ghobadian B, Tavakoli T, Nikbakht A, Najafi G. Application of artificial neural networks for the prediction of performance and exhaust emissions in SI engine using ethanol-gasoline blends. Energy 2010;35(1):65−9.
[68] Tasdemir S, Saritas I, Ciniviz M, Allahverdi N. Artificial neural network and fuzzy expert system comparison for prediction of performance and emission parameters on a gasoline engine. Expert Syst Appl 2011;38(11):13912−23.
[69] Barma SD, Das B, Giri A, Majumder S, Bose P. Back propagation artificial neural network (BPANN) based performance analysis of diesel engine using biodiesel. J Renew Sustain Energy 2011;3(1):013101.
[70] Yap WK, Karri V. Emissions predictive modelling by investigating various neural network models. Expert Syst Appl 2012;39(3):2421−6.
[71] Çay Y, Korkmaz I, Çiçek A, Kara F. Prediction of engine performance and exhaust emissions for gasoline and methanol using artificial neural network. Energy 2013;50:177−86.
[72] Taghavifar H, Khalilarya S, Jafarmadar S. Diesel engine spray characteristics prediction with hybridized artificial neural network optimized by genetic algorithm. Energy 2014;71:656−64.
[73] Rezaei J, Shahbakhti M, Bahri B, Aziz AA. Performance prediction of HCCI engines with oxygenated fuels using artificial neural networks. Appl Energy 2015;138:460−73.
[74] Bietresato M, Calcante A, Mazzetto F. A neural network approach for indirectly estimating farm tractors engine performances. Fuel 2015;143:144−54.
[75] Kapusuz M, Ozcan H, Yamin JA. Research of performance on a spark ignition engine fueled by alcohol−gasoline blends using artificial neural networks. Appl Therm Eng 2015;91:525−34.
[76] Shamekhi A-M, Shamekhi AH. A new approach in improvement of mean value models for spark ignition engines using neural networks. Expert Syst Appl 2015;42(12):5192−218.
[77] Rahimi-Gorji M, Ghajar M, Kakaee A-H, Domiri Ganji D. Modeling of the air conditions effects on the power and fuel consumption of the SI engine using neural networks and regression. J Braz Soc Mech Sci Eng 2017;39(2):375−84.
[78] Bahri B, Shahbakhti M, Aziz AA. Real-time modeling of ringing in HCCI engines using artificial neural networks. Energy 2017;125:509−18.
[79] Tu JV. Advantages and disadvantages of using artificial neural networks versus logistic regression for predicting medical outcomes. J Clin Epidemiol 1996;49(11):1225−31.

[80] Dreiseitl S, Ohno-Machado L. Logistic regression and artificial neural network classification models: a methodology review. J Biomed Inf 2002;35(5−6):352−9.
[81] Togun N, Baysec S. Genetic programming approach to predict torque and brake specific fuel consumption of a gasoline engine. Appl Energy 2010;87(11):3401−8.
[82] Badra J, Sim J, Pei Y, Viollet Y, Pal P, Futterer C, Brenner M, Som S, Farooq A, Chang J. Combustion system optimization of a light-duty GCI engine using CFD and machine learning. SAE Technical Paper; 2020. https://doi.org/10.4271/2020-01-1313.
[83] Badra JA, Khaled F, Tang M, Pei Y, Kodavasal J, Pal P, Owoyele O, Fuetterer C, Mattia B, Aamir F. Engine combustion system optimization using computational fluid dynamics and machine learning: a methodological approach. J Energy Resour Technol 2020;143(2):022306. https://doi.org/10.1115/1.4047978.
[84] Owoyele O, Pal P, Vidal Torreira A. An automated machine learning-genetic algorithm framework with active learning for design optimization. J Energy Resour Technol 2020;143(8):082305. https://doi.org/10.1115/1.4050489.
[85] Owoyele O, Pal P, Torreira AV, Probst D, Shaxted M, Wilde M, Senecal PK. Application of an automated machine learning-genetic algorithm (AutoML-GA) coupled with computational fluid dynamics simulations for rapid engine design optimization. Int J Engine Res 2021. https://doi.org/10.1177/14680874211023466. (In press).
[86] Owoyele O, Pal P. A novel active optimization approach for rapid and efficient design space exploration using ensemble machine learning. J Energy Resour Technol 2020;143(3):032307.
[87] Owoyele O, Pal P. A novel machine learning-based optimization algorithm (ActivO) for accelerating simulation-driven engine design. Appl Energy 2021;285:116455. https://doi.org/10.1016/j.apenergy.2021.116455.
[88] Badra J, Zubail A, Sim J. Numerical investigation into effects of fuel physical properties on GCI engine performance and emissions. Energy Fuels 2019;33(10):10267−81.
[89] Badra J, Viollet Y, Elwardany A, Im HG, Chang J. Physical and chemical effects of low octane gasoline fuels on compression ignition combustion. Appl Energy 2016;183:1197−208.
[90] Cho K, Zhang Y, Cleary D. Investigation of fuel effects on combustion characteristics of partially premixed compression ignition (PPCI) combustion mode at part-load operations. SAE Int J Engines 2018;11(6):1371−84.
[91] Kim D, Martz J, Violi A. Effects of fuel physical properties on direct injection spray and ignition behavior. Fuel 2016;180:481−96.
[92] Som S, Longman D, Ramírez A, Aggarwal S. A comparison of injector flow and spray characteristics of biodiesel with petrodiesel. Fuel 2010;89(12):4014−24.
[93] Mohan B, Yang W, Yu W. Effect of internal nozzle flow and thermo-physical properties on spray characteristics of methyl esters. Appl Energy 2014;129:123−34.
[94] Mohan B, Yang W, Yu W, Tay KL. Numerical analysis of spray characteristics of dimethyl ether and diethyl ether fuel. Appl Energy 2017;185:1403−10.
[95] Pei Y, Shan R, Som S, Lu T, Longman D, Davis MJ. Global sensitivity analysis of a diesel engine simulation with multi-target functions. SAE Technical Paper; 2014.
[96] Popova M, Isayev O, Tropsha A. Deep reinforcement learning for de novo drug design. Sci Adv 2018;4(7):7885.
[97] Gómez-Bombarelli R, Wei JN, Duvenaud D, Hernández-Lobato JM, Sánchez-Lengeling B, Sheberla D, Aguilera-Iparraguirre J, Hirzel TD, Adams RP, Aspuru-Guzik A. Automatic chemical design using a data-driven continuous representation of molecules. ACS Central Sci 2018;4(2):268−76.

[98] Olivecrona M, Blaschke T, Engkvist O, Chen H. Molecular de-novo design through deep reinforcement learning. J Cheminf 2017;9(1):1−14.
[99] Sattarov B, Baskin II, Horvath D, Marcou G, Bjerrum EJ, Varnek A. De novo molecular design by combining deep autoencoder recurrent neural networks with generative topographic mapping. J Chem Inf Model 2019;59(3):1182−96.
[100] Al-Fahemi JH, Albis NA, Gad EA. QSPR models for octane number prediction. J Theoret Chem 2014.
[101] Mendes G, Aleme HG, Barbeira PJ. Determination of octane numbers in gasoline by distillation curves and partial least squares regression. Fuel 2012;97:131−6.
[102] Andrade JM, Muniategui S, Prada D. Prediction of clean octane numbers of catalytic reformed naphthas using FT-mir and PLS. Fuel 1997;76(11):1035−42.
[103] de Paulo JM, Barros JE, Barbeira PJ. A PLS regression model using flame spectroscopy emission for determination of octane numbers in gasoline. Fuel 2016;176:216−21.
[104] Abdul Jameel AG, Elbaz AM, Emwas A-H, Roberts WL, Sarathy SM. Calculation of average molecular parameters, functional groups, and a surrogate molecule for heavy fuel oils using 1H and 13C nuclear magnetic resonance spectroscopy. Energy Fuels 2016;30(5):3894−905.
[105] Flecher PE, Welch WT, Albin S, Cooper JB. Determination of octane numbers and Reid vapor pressure in commercial gasoline using dispersive fiber-optic Raman spectroscopy. Spectrochim Acta Mol Biomol Spectrosc 1997;53(2):199−206.
[106] Guan L, Feng X, Li Z, Lin G. Determination of octane numbers for clean gasoline using dielectric spectroscopy. Fuel 2009;88(8):1453−9.
[107] Abdul Jameel AG, Van Oudenhoven V, Emwas A-H, Sarathy SM. Predicting octane number using nuclear magnetic resonance spectroscopy and artificial neural networks. Energy Fuels 2018;32(5):6309−29.
[108] Smith JS, Nebgen BT, Zubatyuk R, Lubbers N, Devereux C, Barros K, Tretiak S, Isayev O, Roitberg AE. Approaching coupled cluster accuracy with a general-purpose neural network potential through transfer learning. Nat Commun 2019;10(1):1−8.
[109] Yoo P, Sakano M, Desai S, Islam MM, Liao P, Strachan A. Neural network reactive force field for C, H, N, and O systems. Comput Mater 2021;7(1):1−10.
[110] Toyao T, Maeno Z, Takakusagi S, Kamachi T, Takigawa I, Shimizu K-i. Machine learning for catalysis informatics: recent applications and prospects. ACS Catal 2019;10(3):2260−97.
[111] Nareid H, Lightowler N. Detection of engine misfire events using an artificial neural network. SAE Technical Paper; 2004.
[112] Chen SK, Mandal A, Chien L-C, Ortiz-Soto E. Machine learning for misfire detection in a dynamic skip fire engine. SAE Int J Engines 2018;11(6):965−76.
[113] Günnemann N, Pfeffer J. Predicting defective engines using convolutional neural networks on temporal vibration signals. In: Presented at first international workshop on learning with imbalanced domains: theory and applications. PMLR; 2017. p. 92−102. Year. Published.
[114] Yuan M, Wu Y, Lin L. Fault diagnosis and remaining useful life estimation of aero engine using LSTM neural network. In: Presented at 2016 IEEE international conference on aircraft utility systems (AUS). IEEE; 2016. p. 135−40. Year. Published.
[115] Zhu D, Deng J, Wang J, Wang S, Zhang H, Andert J, Li L. Development and application of ion current/cylinder pressure cooperative combustion diagnosis and control system. Energies 2020;13(21):5656.
[116] Kodavasal J, Abdul Moiz A, Ameen M, Som S. Using machine learning to analyze factors determining cycle-to-cycle variation in a spark-ignited gasoline engine. J Energy Resour Technol 2018;140(10).

[117] Kuzhagaliyeva N, Thabet A, Singh E, Ghanem B, Sarathy SM. Using deep neural networks to diagnose engine pre-ignition. Proc Combust Inst 2021;38(4):5915—22.
[118] Schuman CD, Young SR, Mitchell JP, Johnston JT, Rose D, Maldonado BP, Kaul BC. Low size, weight, and power neuromorphic computing to improve combustion engine efficiency. In: Presented at 2020 11th international green and sustainable computing workshops (IGSC). IEEE; 2020. p. 1—8. Year. Published.

SECTION 1

Artificial Intelligence to optimize fuel formulation

CHAPTER 2

Optimization of fuel formulation using adaptive learning and artificial intelligence

Juliane Mueller[1], Namho Kim[2], Simon Lapointe[3], Matthew J. McNenly[3], Magnus Sjöberg[2] and Russell Whitesides[3]

[1]Lawrence Berkeley National Laboratory, Berkeley, CA, United States; [2]Sandia National Laboratories, Livermore, CA, United States; [3]Lawrence Livermore National Laboratory, Livermore, CA, United States

1. Introduction and motivation

Multimode operation couples high-load stoichiometric combustion with a lean and dilute part-load operation. This allows increasing the spark-ignition (SI) engine efficiency at part-load conditions while retaining the high power density and ease of operation of traditional SI engines. Here, the so-called mixed-mode combustion is of interest for the lean and dilute operation. Combustion is initiated by deflagration while the latter is dominated by end-gas autoignition. The mixed-mode combustion allows strong control over combustion parameters via a spark plug and achieves high combustion efficiency by enabling nearly complete combustion as well as high thermal efficiency by maintaining short combustion durations via a controlled end-gas autoignition. To maximize the efficiency gain for multimode engines, optimal fuel compositions must be determined to broaden the operable range of mixed-mode combustion. The search for new fuels that yield optimal performance in new types of combustion systems (e.g., in terms of efficiency) is resource intensive and time consuming when laboratory experiments have to be conducted. Three-dimensional computational fluid dynamics (3D-CFD) simulation models are often used to alleviate this cost by approximating the laboratory observations and to predict which fuels are the most promising to be tried in experimentation. However, detailed simulations are computationally nontrivial as they often require massive amounts of compute time and some underlying physics may not be sufficiently resolved or even understood to be modeled accurately, and thus simplifications are made. Therefore, the usefulness of simulations for finding new fuels is limited especially in optimization settings where many simulations must be run in order to find the best fuel candidates.

This study focuses on using advanced optimization methods and machine learning to find the best fuels for multimode engine operation that is simulated using a high-fidelity chemistry mechanism with a low-cost fluid dynamics model. The simple fluid dynamics model parameters can be trained either from higher-fidelity CFD or directly from engine experiments.

In order to alleviate the computational burden of 3D-CFD calculations, a combination of zero-dimensional (0D) and one-dimensional (1D) models is used. These lower-fidelity models lead to faster-to-evaluate simulations, which are more suitable to be queried in an iterative optimization scheme for finding optimal fuels and engine operation. We propose to couple advanced derivative-free optimization techniques with these low-fidelity simulation models to determine the best fuels to be used in mixed-mode combustion. Surrogate model–based optimization that employs Gaussian process (GP) models and adaptive sampling of the fuel search space is used to minimize the required number of simulation model calls. This approach is compared to the performance of a genetic algorithm (GA). In order to satisfy established requirements on the Research Octane Number (RON) for feasible mixed-mode operation, a neural network (NN) has been developed to predict a fuel composition's RON. The automated optimization approaches, and, in particular, the GP-based approach to fuel search allow to quickly find the most promising fuels, and thus have the potential to close the feedback loop between data collection by laboratory experimentation, simulation validation, and identification of optimal fuels.

This chapter is organized as follows. In Section 2, details are provided for the mixed-mode combustion simulations and for the performance metrics (cost functions) used in the fuel search. A constraint on the fuel's RON is imposed in the optimization and the NN used to predict the RON is described in Section 3. The optimization problem that aims at maximizing the performance metrics is formulated in Section 4 and two optimization algorithms for solving the problem are presented. Section 5 contains the description of our numerical experiments and the outcomes. The results are discussed in Section 6 and Section 7 summarizes and concludes the chapter.

2. Mixed-mode combustion and fuel performance metrics

Mixed-mode combustion uses deflagration to consume approximately the first half of the charge and controlled end-gas autoignition to quickly consume the rest of the charge. This helps to maintain a sufficiently short combustion duration which is critical to exploit the high thermal efficiency benefit from lean and dilute combustion [43]. Because the mixed-mode combustion relies on controlled end-gas autoignition, combustion characteristics are very sensitive to various factors including the fuel properties. Thus, to maximize the operable range and efficiency gain from mixed-mode combustion, it is crucial to find not only the technical implementations to achieve stable initial deflagration but also fuels that allow for robust control over end-gas autoignition.

Since laboratory experiments and high-fidelity multidimensional simulations are too time-consuming to be used within fuel optimization studies, a dedicated framework of 0D kinetic simulation and 1D engine simulation was created. This framework relies on an adaptively preconditioned 0D homogenous chemical mechanism integrator to resolve the end-gas autoignition process [41], but taking initial and boundary conditions from 1D simulations to model the system with an appropriate level of fidelity. Previous work reports that the end-gas autoignition process in stoichiometric SI engines can be modeled with sufficient fidelity with simplicity by imposing an experimentally measured cylinder pressure trace in a 0D reactor model [44]. This methodology provides the basis for the framework introduced in this study, with a difference that the imposed cylinder pressure trace was derived from a 1D engine simulation.

In this work, the Zero-RK software package [41] (see also https://github.com/LLNL/zero-rk) with a kinetic mechanism containing more than 2800 species and 12,800 reactions [42] is used to predict the end-gas autoignition of the fuel mixtures considered. The reactor model simulates dynamic changes in temperature and chemical species concentrations according to the imposed pressure boundary, capturing the fuel chemistry leading to autoignition with great detail. For 1D engine simulations to generate initial and boundary conditions for 0D simulations, the GT-Power software package was utilized. The in-cylinder combustion rate in the GT-Power simulations was prescribed using multiple Wiebe functions. The parameters used in the Wiebe functions were determined based on past experiments [45].

The GT-Power model of the experimental engine was developed and validated over a wide range of conditions. Then the developed model was exercised to alter and predict in-cylinder thermodynamic conditions by spanning engine speeds of 1000, 1400, and 2000 RPM, intake pressures of 1 and 1.3 bar, Bottom Dead Center (BDC) temperatures of 85, 100, 115, and 130°C, equivalence ratios of 0.45 and 0.5, and compression ratios of 12, 13, and 14. Unlike in the experiments where adjusting BDC temperature is done by changing intake air temperature, variable valve timing and lift feature were used to change the amount of trapped residual. The results of these 1D simulations were used to impose the initial conditions and the time-varying pressure profile in a 0D reactor model.

The 0D reactor model then computes the ignition delay time (IDT), which then allows one to determine the crank angle as well as the accumulated heat release at ignition. Based on the experimentally informed operating region (see Fig. 2.1), the information on the crank angle and the amount of end-gas mass at autoignition timing are used to determine if a given fuel can be operated with mixed-mode combustion at a given condition without issues with reactivity of the fuel mixture. If the fuel mixture is too reactive, it is likely to incur unacceptably strong acoustic knock. On the other hand, if the mixture is not reactive enough, end-gas mass is unlikely to autoignite before the charge is consumed by deflagration, failing to realize mixed-mode combustion.

In order to assess a fuel's performance, two measures were developed, namely a load range criterion that reflects the difference between the maximum and the minimum engine loads for the points that fall within the allowable operating region (points inside the black polygon in Fig. 2.1) and a robustness criterion that reflects the number of operable points that fall within the maximum and the minimum engine loads. These values were computed for three different engine speeds (1000, 1400, and 2000 RPM) at

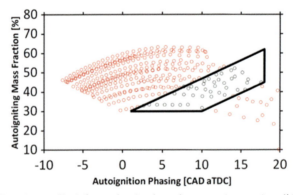

Figure 2.1 Experimentally informed mixed-mode operating region (*black polygon*) and operating points of a given fuel. Only the operable points that fall within the polygon (combination of fuel and engine operation conditions) are able to realize mixed-mode combustion that is both stable and knock-free.

compression ratio 12, and the three robustness and load range values, respectively, were added up to reflect the fuel's performance, which represent our cost functions to be maximized.

3. A neural network model to predict fuel research octane numbers

An investigation into what type of fuels offers the greatest range of mixed-mode operation showed that it has a desire for high-RON, high-octane-sensitivity fuels, similar to modern SI engines, allowing us to reduce the size of the fuel search space.

Performance of spark-ignited fuels is highly dependent on the antiknock characteristics which are commonly quantified by the research and motor octane numbers (RON and MON) [1].

There is an extensive body of research in estimation of octane numbers based on mixture composition, average chemical structure, and simulated ignition propensity [2–12]. Prediction root-mean-square-errors (RMSE) greater than 2 ON with worst case error greater than 10 ON are typical for these types of models. In this work, a new methodology for octane prediction is used that leverages both detailed chemical kinetics for ignition calculations and chemical structure information to achieve high accuracy predictions for RON and MON. In addition to accuracy, a key objective is that the prediction method be computationally cheap enough to be included in fuel optimization loops [3,13].

For the current study, data were collected from the published literature for ASTM D2699 and D2700 measurements of over 500 different gasoline surrogates and pure components [4,5,10,14–37]. In total, 45 different fuel molecules are represented in the data set including common reference molecules and promising fuel components and additives. These data are used to train a feed-forward NN [38] with the task of predicting octane number as a function of constant volume IDT and average molecular structure information. The ignition inputs to the NN are inverse IDT at 750 K for RON (825 K for MON) and the slope of the IDT curve with respect to temperature at 850 K (both RON and MON). The temperatures for each IDT were selected based on mutual information [39] between the IDT at temperatures between 700 and 900 K and the target quantities, i.e., RON and MON. The slope of the IDT curve has been previously shown to correlate with fuel sensitivity [6]. The molecular structure information input to the NN is the number of hydrogen atoms and seven atom types defined based on atomic number, number of covalent bonds, hybridization, and whether the atom is part of a ring structure. For mixtures, the structure information is computed as the mole averaged contribution from each mixture component. All the inputs for the NN are transformed by subtracting the mean and dividing by the standard deviation to aid in NN training.

During NN training, L_2 regularization was added as a penalty on the NN weights to prevent overfitting. The Adam optimization algorithm was used with the RMSE between measured and predicted octane numbers as the objective function. The NN architecture [number of layers (1), number of nodes per layer (24), activation function (ReLU), learning rate (0.005), and L_2 penalty factor (1.0e−4)] were selected by grid search using fivefold cross-validation. The NN was implemented in the TensorFlow framework [40].

Fig. 2.2 shows the error in prediction over the data set for the RON (left) and MON (right) models. The resulting RMSE of prediction for the RON model was 0.938 (R^2 value of 0.996) and for the MON model it was 0.979 (R^2 value of 0.992). To further test the predictive capability of the models, they were tested on measurements from mixtures that were not available when the model was created. A set of 19 surrogate fuels with widely varying chemical composition was used as a hold-out validation set. The overall RMSE in prediction for this validation set was 0.85 for RON and 1.6 for MON, providing confidence in the quality of predictions using these models.

4. Optimization problem formulation and description of solution approaches

4.1 Constrained optimization formulation

We formulate the fuel search task as an optimization problem in which we search over the fuel components $x = [x_1, ..., x_9]$, $x_i \in [0, 1], i = 1, ..., 9$.

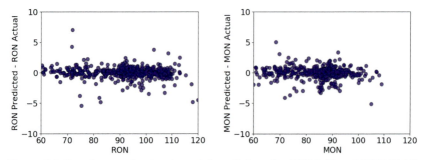

Figure 2.2 Error in neural network model predictions for RON (left) and MON (right).

When mixed, the resulting fuel composition, denoted by $m(x)$, must satisfy constraints on RON, and the goal is to maximize a cost function:

$$\max_{x \in [0,1]^9} f(m(x)) \tag{2.1}$$

$$95 \leq RON(m(x)) \leq 98 \tag{2.2}$$

$$\sum_{i=1}^{9} x_i = 1 \tag{2.3}$$

Here, the cost function $f(\cdot)$ in Eq. (2.1) is either the robustness measure or the load range measure of the fuel composition introduced in Section 2. It is evaluated with the zero-RK kinetic simulation following the GT-Power computed pressure-temperature trajectory, which, depending on how many processors are available, may take from 0.5 to 2.5 min for each fuel composition. The RON constraint function in Eq. (2.2) is evaluated with the NN model introduced in Section 3, and it takes about 5 s to evaluate. Being composition fractions, the fuel components have to sum up to 1 (Eq. 2.3).

Solving this optimization problem is nontrivial as both the objective and the constraint functions are black boxes, i.e., we do not have an analytic description of the functions and derivative information is not available. Therefore, derivative-free optimization methods are good options [50,51]. Due to the computational expense associated with the objective and the constraint function evaluations, the goal is to try as few fuel mixtures as possible during the search in order to keep the optimization time tractable. Since the objective function evaluation is independent of the RON constraint and the RON constraint is significantly faster to evaluate, the optimizer should first evaluate the RON constraint and only if it is satisfied should the objective function be evaluated. A fuel mixture that does not satisfy the RON constraint will not be a solution to the problem, and thus no computation time should be wasted by evaluating the objective function at these infeasible solutions. This will save significant amounts of compute time and lead to results faster. Two different types of derivative-free algorithms that employ this strategy are compared, namely a GA and a GP model–based algorithm.

4.2 Genetic algorithm

A GA [46] is a search algorithm inspired by the process of natural selection. The algorithm uses a pool, called population, of possible solutions to the optimization problem. Each solution is called an individual and these individuals undergo mutations and combinations (also called crossover).

Each individual is assigned a fitness value based on the objective function of the optimization problem. Individuals with a higher fitness value are given a higher mating (combination) probability. The combinations produce "children," and the process is repeated over many generations, evolving better solutions until a stopping criterion is reached. A flowchart of the GA is illustrated in Fig. 2.3.

GAs are stochastic since the initial population is chosen randomly and some stochasticity is present in the mutation, crossover, and parent selection processes. For this reason, there is no guarantee on the optimality of the final solution. However, GAs do not require derivative information and are well-suited for problems with multimodal cost function landscapes for which gradient-based methods typically stop in local optima.

In the present fuel search optimization problem, the fitness function to be maximized is defined as the fuel's load range (or robustness). A fitness value of zero is returned for fuels with a RON outside the desired range (see Eq. 2.2) as predicted by the NN. A GA with a population size of 300 is used. The population is evolved over 10 generations, with a mutation probability of 20% and a crossover probability of 75%. The large population size and high mutation probability help achieve a good exploration of the solution (fuel) space. Furthermore, multiple realizations of the GA search are conducted with different initial populations to mitigate the impact of the stochasticity associated with the population initialization.

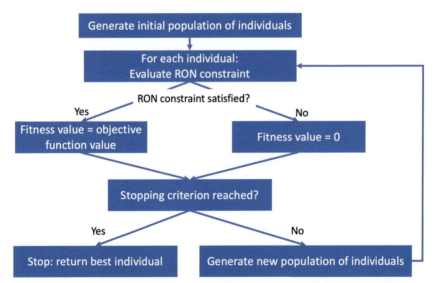

Figure 2.3 Flowchart of the genetic algorithm (GA) used for the mixed-mode fuel optimization task.

4.3 Gaussian process—based surrogate model optimization algorithm

The second optimization method makes use of GP models [47]. GPs are computationally cheap surrogate models that are widely used in black-box optimization algorithms to approximate the expensive-to-evaluate simulations and to guide the adaptive selection of sample points [48]. The GP is trained on input—output data pairs (e.g., for the objective function $\{(x_j, f(x_j))\}_{j=1}^{n}$, where n denotes the number of input-output pairs obtained so far) and used for making predictions at inputs for which the output has not yet been computed. In addition to the prediction of the output value, an uncertainty estimate is obtained. Both the predicted function values and their uncertainty estimates are then used to explore the search space and select new fuels for evaluation, which is done by balancing global ("exploration") and local ("exploitation") search.

For the current optimization problem, two separate GPs are used, namely one that approximates the objective function and one that approximates the constraint function. The GP model—assisted algorithm begins by creating an initial experimental design which consists of n_0 fuel compositions. By definition, the sample points in the initial design satisfy the equality constraint (Eq. 2.3). For each point in the initial design, the RON constraint (Eq. 2.2) is evaluated. If the RON constraint is not satisfied at a given sample point, the objective function is not evaluated. Given the values of the constraint and the objective function, two different GP models are built. The GP for the objective function is likely based on fewer points than the GP for the constraint since the objective function is not evaluated at infeasible points. Using the GPs, an auxiliary optimization problem is formulated (Eqs. (2.4)—(2.7)) and solved, and its solution forms the next sample point for which the RON constraint is evaluated. The algorithm iterates between constraint function (and objective function, if applicable) evaluation, updating of the GP models, and sample point selection until a predefined budget of function evaluations has been reached. The flowchart of the GP model—assisted algorithm is illustrated in Fig. 2.4.

The auxiliary optimization problem is defined as follows:

$$\max_{x \in [0,1]^9} \mathbb{E}I(x) \tag{2.4}$$

$$\mathbb{E}V_l(x) \leq \delta_n \tag{2.5}$$

$$\mathbb{E}V_u(x) \leq \delta_n \tag{2.6}$$

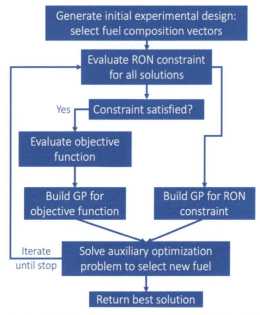

Figure 2.4 Flowchart of the Gaussian process model–guided optimization method.

$$\sum_{i=1}^{9} x_i = 1 \qquad (2.7)$$

Here, $\mathbb{E}I(x)$ in Eq. (2.4) denotes the expected improvement at a point x. It is calculated based on the GP model's prediction of the objective function $(\widehat{f}(x))$ and the corresponding uncertainty estimate $(\widehat{s}_f(x))$:

$$\mathbb{E}I(x) = \begin{cases} \left(\widehat{f}(x) - f_{max}\right) \times \Phi\left(\dfrac{\widehat{f}(x) - f_{max}}{\widehat{s}_f(x)}\right) + \widehat{s}_f(x) \times \phi\left(\dfrac{\widehat{f}(x) - f_{max}}{\widehat{s}_f(x)}\right), & \text{if } x \notin C \text{ and } \widehat{s}_f(x) > 0 \\ 0, & \text{if } x \in C \text{ or } \widehat{s}_f(x) = 0 \end{cases}$$

Here, Φ and ϕ denote the cumulative and probability distribution functions of the standard normal distribution, respectively. The expected improvement is 0 at all points x in the set C at which the constraint has already been evaluated. At these points, either the objective function value is already known, or the constraint function was violated, and thus no improvement is expected. The value f_{max} is the best function value found so far.

$\mathbb{E}V_l(\boldsymbol{x})$ and $\mathbb{E}V_u(\boldsymbol{x})$ in Eqs. (2.5) and (2.6), respectively, denote the expected violation of the lower and upper bounds on the RON constraint:

$$\mathbb{E}V_l(\boldsymbol{x}) = \begin{cases} \left(l - \widehat{g}(\boldsymbol{x})\right) \times \Phi\left(\dfrac{l - \widehat{g}(\boldsymbol{x})}{\widehat{s}_g(\boldsymbol{x})}\right) + \widehat{s}_g(\boldsymbol{x}) \times \phi\left(\dfrac{l - \widehat{g}(\boldsymbol{x})}{\widehat{s}_g(\boldsymbol{x})}\right), & \text{if } \widehat{s}_g(\boldsymbol{x}) > 0 \\ 0, & \text{if } \widehat{s}_g(\boldsymbol{x}) = 0 \end{cases}$$

$$\mathbb{E}V_u(\boldsymbol{x}) = \begin{cases} \left(\widehat{g}(\boldsymbol{x}) - u\right) \times \Phi\left(\dfrac{\widehat{g}(\boldsymbol{x}) - u}{\widehat{s}_g(\boldsymbol{x})}\right) + \widehat{s}_g(\boldsymbol{x}) \times \phi\left(\dfrac{\widehat{g}(\boldsymbol{x}) - u}{\widehat{s}_g(\boldsymbol{x})}\right), & \text{if } \widehat{s}_g(\boldsymbol{x}) > 0 \\ 0, & \text{if } \widehat{s}_g(\boldsymbol{x}) = 0 \end{cases}$$

where $\widehat{g}(\boldsymbol{x})$ is the predicted RON constraint function value at $\boldsymbol{x}$, $\widehat{s}_g(\boldsymbol{x})$ is the corresponding uncertainty, and l and u are the lower and upper bounds on the RON constraint, respectively (see Eq. 2.2). The expected violation is zero at points where the constraint function has already been evaluated (the uncertainty is zero), otherwise it is larger than zero. The value for δ_n (the upper limit on the allowed expected violation) is adjusted dynamically and depends on the number n of objective function evaluations done so far:

$$\delta_n = \frac{1 - \dfrac{\log(n - n_0 + 1)}{\log(n_{max} - n_0)}}{2(n - n_0 + 1)}$$

where n_0 denotes the number of points in the initial experimental design and n_{max} is the maximum number of objective function evaluations that is allowed. Thus, the value for the limit δ_n decreases as the algorithm iterates, and thus the algorithm is forced to hone in on points that are more and more likely to satisfy the RON constraint (see Fig. 2.5).

5. Numerical experiments and results

Numerical experiments were carried out with both optimization algorithms for a search over nine different fuel components from five fuel groups (paraffins, isoparaffins, olefins, naphthenes, aromatics). The fuel components considered were toluene, 1,2,4-trimethylbenzene, iso-octane, n-heptane, iso-pentane, n-pentane, 1-hexene, di-isobutylene, and cyclopentane. Two separate optimization problems were solved: (1) maximize the fuel mixture's robustness and (2) maximize the fuel mixture's load range. Since both optimization algorithms have stochastic components, five trials were run with each algorithm for each problem.

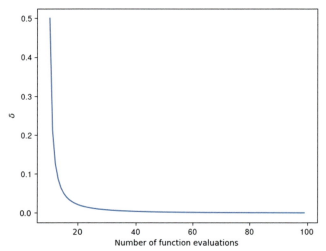

Figure 2.5 δ_n as a function of the number of evaluations done so far. Here, $n_0 = 10$, $n_{max} = 100$. As the budget of allowable function evaluations is approached, the allowed expected constraint violation approaches zero.

The GA used a population of 300 individuals that were evolved over 10 generations, i.e., a total of 3000 fuel compositions were tried. The GP-assisted algorithm used $n_{max} = 300$ function evaluations. Fig. 2.6 shows the convergence plots for both algorithms when maximizing the load range (left) and robustness (right), respectively.

For the load range, we observe that after about 25 function evaluations, the GP-based approach consistently finds better solutions with less

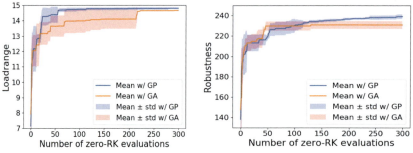

Figure 2.6 Convergence plot for maximizing the load range (*left*) and robustness (*right*) versus the number of objective function evaluations (higher graphs are better). Shown are the results for the genetic algorithm (GA) and the Gaussian Process (GPs) approach. The solid lines represent the mean performance over five trials and the shaded areas represent the standard deviations (narrower standard deviations are better).

variability (narrower standard deviations). For the robustness, the GP outperforms the GA after approximately 125 evaluations and has smaller standard deviations. Note that shown are only the first 300 randomly generated individuals of the GA, i.e., the first generation. At this point, no evolution/optimization has taken place yet. However, even after evolving the population over 10 generations (3000 evaluations), the solutions found by the GP after 300 evaluations are still better: for the robustness measure, the GA achieves after 3000 evaluations an average (standard deviation) of 235.675 (1.096) whereas the GP reaches after 300 evaluations 239.436 (1.984); and for the load range, the GA reaches after 3000 evaluations a mean (standard deviation) of 14.776 (0.055) whereas the GP achieves after 300 evaluations 14.802 (0.057). Especially when computationally expensive simulations must be queried during the optimization, being able to find good solutions within a very limited number of queries to the expensive function is important. On the other hand, when function evaluations are computationally inexpensive and thousands of queries can be made, using a GA may be more appropriate as the computational overhead of the GA itself is significantly less than that of the GP.

Fig. 2.7 shows the optimal fractions of fuel groups found by the GP algorithm that maximize the load range. Shown are also the optimal load range values, and it can be observed that they all lie within 1% of each other. Since different fuel compositions lead to approximately the same performance in terms of load range, it can be concluded that the objective function landscape has multiple local optima. Olefins and aromatics have the highest fractions, and paraffins are present in the solutions of all five

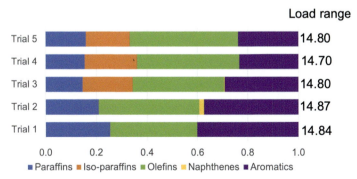

Figure 2.7 Optimal fractions of fuel groups found by the Gaussian process (GP) algorithm that maximize the load range (here shown as the sum for three engine speeds and given as indicated mean effective pressure (IMEP) in units of [kPa]. The optimal load range values (*right side*) lie within 1% of each other.

trials. Only very little naphthenes are included in the mixture, and iso-paraffins are present in three of the five solutions, indicating that it is possible to achieve optimal performance without them.

Also for the robustness, it can be observed that multiple different compositions lead to approximately the same performance (fuel group fractions shown in Fig. 2.8) and generally, very little isoparaffins are used and no naphthenes are present in the optimal mixtures.

A noticeable difference between the mixtures that maximize the load range and the robustness is that for achieving highest robustness, the amount of olefins is significantly larger in four out of five trials than when maximizing the load range (more than 80% of olefins). The one trial in which less olefins were used has the worst robustness value out of the five trials, and instead the mixture contains more isoparaffins and aromatics.

When looking into the detailed fuel compositions for maximizing the load range, four out of five trials consist of 35% or more of di-isobutylene. The remaining trial contains about 30% di-isobutylene and 40% trimethylbenzene, which constitutes less than 15% in the other trials. Cyclopentane and 1-hexene have almost no contribution to the fuel mixture. When maximizing the robustness, di-isobutylene is the largest fuel component in the mixture for all five trials, whereas no cyclopentane and only a very small fraction of n-pentane, n-heptane, and iso-octane were in the optimal mixtures.

6. Discussion

The results of our numerical study show that the GP-based algorithm is able to find better-performing fuel compositions within fewer function evaluations than the GA. After 300 evaluations, the GP reached better

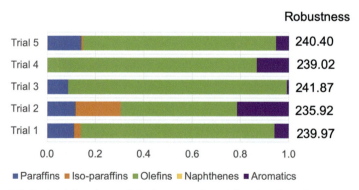

Figure 2.8 Optimal fractions of fuel groups found by the Gaussian process (GP) algorithm that maximize the robustness. The optimal robustness values (*right side*) lie within 2% of each other.

robustness and load range values that have over five trials less variability than the GA after 3000 evaluations. This is particularly an important feature when function evaluations are computationally intensive, i.e., when good solutions must be found within very few queries to the expensive simulation. Although the amount of time required for evaluating the objective function has already been significantly reduced by using 1D and 0D approximations of 3D-CFD models, the computational overhead of evaluating the load range and robustness for a specific fuel mixture is still nontrivial. Therefore, surrogate models that adaptively learn the objective function landscape and sample the search space are an attractive optimizer option.

On the other hand, when function evaluations are computationally cheap (e.g., fractions of a second), or if enough compute power is available to do hundreds of function evaluations in parallel, the GA might be the better optimization option. For example, if one could do 100 evaluations in parallel, then the population could be evolved over 300 generations, and thus 30,000 evaluations could be done within approximately the same wall clock time in which the GP performs 300 consecutive evaluations. There are also opportunities to select more than one new sample point in each iteration of the GP algorithm. For example, when maximizing the auxiliary optimization problem, the various local optima of the auxiliary objective function could be selected as new evaluation points, in which case the GP is updated with multiple new points in each iteration. This, however, does not change the fact that the GP's own computational overhead (for fitting the GP to the input-output data pairs) is significantly larger than that of the GA. Therefore, optimizers that use surrogate models such as the one used here should be reserved for problems with costly function evaluations.

The auxiliary optimization problem solved in each iteration of the GP algorithm could also be formulated differently. For example, one could add the RON constraint to the expected improvement objective function using a penalty term and solve an unconstrained optimization problem. Unconstrained optimization problems are usually easier to solve, but the adjustment of the penalty factor is not straight-forward and the final solution is not necessarily feasible. One could also formulate an auxiliary optimization problem in which the expected improvement is multiplied by a probability of constraint violation. Also this would result in an unconstrained problem. However, when multiple constraints are present, the auxiliary objective function is likely to be flat and thus hard to optimize.

The optimization formulation presented here does not take into account any composition limits that may be placed on fuel producers or importers. For example, the California Air Resources Board [49] currently limits the olefin content of reformulated gasoline to 6% and the aromatic content to 25% by volume. While all of the optimal blends presented here could not be sold in California, these specific composition constraints could be easily added to the problem formulation and the GP algorithm would still deliver significant performance gains over the GA method. Other composition limits could also be added based on the boiling point of each component to ensure that the distillation properties of the final blend meet a given specification. Moreover, the described optimization methodology can straight-forwardly be extended to more fuel components such as biofuels by simply adding them to the composition vector x. For example, the current Co-Optima mechanism [42] has 17 bio-derived components that can be added to the composition vector. These biofuels include:

1. Methanol
2. Ethanol
3. 1-propanol
4. 2-propanol
5. 1-butanol
6. 2-methyl-1-propanol
7. 1-pentanol
8. Di-isobutylene (already in palette)
9. 2-methylfuran
10. 2,5-dimethylfuran
11. Anisole
12. 2-butanone
13. Methylbutanoate
14. Methylacetate
15. Ethylacetate
16. Dimethylether
17. Prenol
18. Isoprenol

7. Summary and concluding remarks

In this chapter, two approaches were described for automatically searching over the space of fuel compositions with the goal to maximize the performance of mixed-mode combustion systems. Both methods are derivative-free optimizers that are able to directly take into account the constraints imposed on the fuel's RON. One method is based on GAs and the other adaptively learns the objective function landscape using GP models as approximations of the compute-intensive functions. Two optimization formulations were used, namely one that aims to maximize the fuel composition's load range and another that maximizes the fuel composition's robustness. The results showed that the GP-based optimizer is able to find better fuel compositions within fewer function evaluations than the GA, making it the go-to optimizer for problems that involve computationally expensive functions and that do not allow for thousands of function queries. The numerical study showed that different trials with the optimizer lead to different optimal fuel compositions with approximately the same performance, indicating that some fuel components can be traded off without losing performance. The presented optimization framework is general and flexible enough to be extended to problems with larger fuel search spaces, for example, including biofuels, as well as additional constraints to ensure local restrictions on allowable fuel contents are satisfied.

Acknowledgments

This research was conducted as part of the Co-Optima Initiative sponsored by the U.S. DOE EERE, Bioenergy Technologies and Vehicle Technologies Offices under contract number DE-AC02-05CH11231 (Lawrence Berkeley National Laboratory), under Contract DE-AC52-07NA27344 (Lawrence Livermore National Laboratory), and DE-NA0003525 (Sandia National Laboratories). Co-Optima is a collaborative initiative of multiple national laboratories initiated to simultaneously accelerate the introduction of affordable, scalable, and sustainable biofuels and high-efficiency, low-emission vehicle engines. The authors gratefully acknowledge the support and direction of Alicia Lindauer at BETO, Kevin Stork at VTO, and the Co-Optima leadership team. Sandia National Laboratories is a multimission laboratory managed and operated by National Technology and Engineering Solutions of Sandia, LLC., a wholly owned subsidiary of Honeywell International, Inc., for the U.S. Department of Energy's National Nuclear Security Administration.

References

[1] Section 5. Vol. 05.01 Annual book of ASTM standards 2013: C1234 - D3710. Petroleum products, lubricants, and fossil fuels. Petroleum products and lubricants (I). ASTM International; 2013.
[2] Knop V, Loos M, Pera C, Jeuland N. Fuel 2014;115:666—73.
[3] Ahmed A, Goteng G, Shankar VSB, Al-Qurashi K, Roberts WL, Sarathy SM. Fuel 2015;143:290—300.
[4] AlRamadan AS, Sarathy SM, Khurshid M, Badra J. Fuel 2016;180:175—86.
[5] Yuan H, Yang Y, Brear MJ, Foong TM, Anderson JE. Fuel 2017;188:408—17.
[6] Mehl M, Chen JY, Pitz WJ, Sarathy SM, Westbrook CK. Energy Fuels 2011; 25:5215—23.
[7] Singh E, Badra J, Mehl M, Sarathy SM. Energy Fuels 2017;31:1945—60.
[8] Naser N, Yang SY, Kalghatgi G, Chung SH. Fuel 2017;187:117—27.
[9] Naser N, Sarathy SM, Chung SH. Combust Flame 2018;188:307—23.
[10] Badra JA, Bokhumseen N, Mulla N, Sarathy SM, Farooq A, Kalghatgi G, Gaillard P. Fuel 2015;160:458—69.
[11] Westbrook CK, Sjöberg M, Cernansky NP. Combust Flame 2018;195:50—62.
[12] Corrubia JA, Capece JM, Cernansky NP, Miller DL, Durrett RP, Najt PM. Combust Flame 2020;219:359—72.
[13] Sarathy SM, Kukkadapu G, Mehl M, Javed T, Ahmed A, Naser N, Tekawade A, Kosiba G, AlAbbad M, Singh E, Park S, Rashidi MA, Chung SH, Roberts WL, Oehlschlaeger MA, Sung C-J, Farooq A. Combust Flame 2016;169:171—93.
[14] AlRamadan AS, Badra J, Javed T, Al-Abbad M, Bokhumseen N, Gaillard P, Babiker H, Farooq A, Sarathy SM. Combust Flame 2015;162:3971—9.

[15] J.E. Anderson, T.G. Leone, M.H. Shelby, T.J. Wallington, J.J. Bizub, M. Foster, M.G. Lynskey, D. Polovina. Octane numbers of ethanol-gasoline blends: measurements and novel estimation method from molar composition. SAE Tech. Paper 2012. https://doi.org/10.4271/2012-01-1274. 2012-01−1274.
[16] Andrae JCG, Head RA. Combust Flame 2009;156:842−51.
[17] H.S. Aronsson, M. Tuner, B. Johansson, in: SAE Technical Papers. 2014;1, pp. 2014-01-1303.
[18] Badra J, AlRamadan AS, Sarathy SM. Appl Energy 2017;203:778−93.
[19] Boehman AL, Luecke J, Fouts L, Ratcliff M, Zigler BT, McCormick R. Ignition delay measurements of four component model gasolines exploring the impacts of biofuels and aromatics. Proc Combust Inst 2021;48(4):5549−55.
[20] Cancino LR, da Silva A, De Toni AR, Fikri M, Oliveira AAM, Schulz C, Curran HJ. Fuel 2020;261:116439.
[21] Christensen E, Yanowitz J, Ratcliff M, McCormick RL. Energy Fuels 2011;25:4723−33.
[22] da Silva Jr A, Hauber J, Cancino LR, Huber K. Fuel 2019;243:306−13.
[23] Scott EJY. In: Proc. API Div. Refin.; 1958.
[24] Fan Y, Duan Y, Han D, Qiao X, Huang Z. Int J Engine Res 2019;22(1):39−49.
[25] Foong TM, Morganti KJ, Brear MJ, da Silva G, Yang Y, Dryer FL. Fuel 2014;115:727−39.
[26] Herzler J, Fikri M, Hitzbleck K, Starke R, Schulz C, Roth P, Kalghatgi GT. Combust Flame 2007;149:25−31.
[27] Modification of CFR test engine unit to determine octane numbers of pure alcohols and gasoline-alcohol blends. In: Hunwartzen I, editor. SAE Technical Paper; 1982. p. 820002.
[28] Javed T, Lee C, AlAbbad M, Djebbi K, Beshir M, Badra J, Curran H, Farooq A. Combust Flame 2016;171:223−33.
[29] Kalaskar V, Kang D, Boehman AL. Energy Fuels 2017;31:11315−27.
[30] Kalghatgi G, Babiker H, Badra J. SAE Int J Engines 2015;8:505−19.
[31] Kalghatgi G, Risberg P, Angstrom H-E. In: SAE tech. Pap. SAE International; 2003.
[32] G.T. Kalghatgi, K. Nakata, K. Mogi, in: 2005, pp. 2005-01−0244.
[33] McCormick RL, Fioroni G, Fouts L, Christensen E, Yanowitz J, Polikarpov E, Albrecht K, Gaspar DJ, Gladden J, George A. SAE Int J Fuels Lubr 2017;10:442−60.
[34] Morgan N, Smallbone A, Bhave A, Kraft M, Cracknell R, Kalghatgi G. Combust Flame 2010;157:1122−31.
[35] Morganti KJ, Foong TM, Brear MJ, da Silva G, Yang Y, Dryer FL. Fuel 2013;108:797−811.
[36] H. Solaka, M. Tuner, B. Johansson, SAE Technical Paper 2013. https://doi.org/10.4271/2013-01-0903. 2013-01-0903.
[37] Yuan H, Chen Z, Zhou Z, Yang Y, Brear MJ, Anderson JE. Fuel 2020;261:116243.
[38] Haykin SS. Neural networks: a comprehensive foundation. Prentice Hall; 1999.
[39] Kraskov A, Stögbauer H, Grassberger P. Phys Rev E 2004;69:066138.
[40] Abadi M, et al. TensorFlow: large-scale machine learning on heterogeneous systems. 2015.
[41] McNenly MJ, Whitesides RA, Flowers DL. Proc Combust Inst 2015;35:581−7.
[42] Mehl M, Zhang K, Wagnon SW, Kukkadapu G, Westbrook CK, Pitz WJ, Zhang Y, Curran HJ, Al Rachidi M, Atef N, Sarathy SM. In: 10th US national combustion meeting; 2017. Paper 1A17.
[43] Ayala F, Heywood J. SAE Technical Paper 2007. https://doi.org/10.4271/2007-24-0030. 2007-24-0030.
[44] Kim N, Vuilleumier D, Sjöberg M, Yokoo N, et al. SAE Int J Adv Curr Prac Mobility 2019;1(4):1560−80. https://doi.org/10.4271/2019-01-1140.

[45] Hu Z, Zhang J, Sjöberg M, Zeng W. Int J Engine Res 2020;21(9):1678−95. https://doi.org/10.1177/1468087419889702.
[46] Mitchell M. An introduction to genetic algorithms. Cambridge, MA: MIT Press; 1996.
[47] Cressie NA. Math Geol 1990;22(3):239−52.
[48] Jones DR, Schonlau M, Welch WJ. J Global Optim 1998;13:455−92.
[49] https://ww2.arb.ca.gov/sites/default/files/2020-05/California_Reformulated_Gasoline_Regulations_2-16-14.pdf p.70.
[50] Mueller J, Day M. Inf J Comput 2019;31(4):689−702.
[51] Mueller J, Woodbury JD. J Global Optim 2017;69:117−36. https://doi.org/10.1007/s10898-017-0496-y.

CHAPTER 3

Artificial intelligence—enabled fuel design

Kiran K. Yalamanchi[1], Andre Nicolle[2] and S. Mani Sarathy[1]

[1]Clean Combustion Research Center (CCRC), King Abdullah University of Science and Technology, Thuwal, Western Province, Saudi Arabia; [2]Aramco Fuel Research Center, Aramco Overseas, Rueil-Malmaison, Paris, France

1. Transportation fuels

1.1 Fuel representation

Transportation fuels are complex mixtures resulting from the blending of several streams involved in separation, conversion, upgrading, and blending processes. Many properties of fuels are regulated to ensure a safe and adequate behavior during storage, blending, distribution, and end-use. These properties range from fundamental macroscopic physicochemical data (density, viscosity) which can be readily predicted using first-principle physical models to more complex engineering indicators such as flash point (FP) or octane number (ON) for which artificial intelligence (AI) techniques may outperform traditional bottom-up physical modeling [1] at the expense of traditional chemical insight [2].

For most practical fuels, detailed chemical analysis is unavailable and only its PIONA (paraffinic, iso-paraffinic, olefinic, naphthenic, aromatic) composition is provided. Fuel formulation methodologies based on detailed composition, therefore, require a molecular reconstruction step. Deep learning (DL) approaches were recently shown to outperform classical fuel reconstruction method such as entropy maximization [3].

A more global fuel representation would lie in the formulation of physicochemical surrogates, i.e., mixtures of a few pure species (typically less than three per chemical family) emulating real fuel properties [4] to enable fundamental well-controlled experiments on simplified multicomponent mixtures. An attempt to design a jet fuel surrogate through multiple linear regression (MLR) and support vector machine (SVM) models trained on semiquantitative [5] reactive force field simulations has been recently reported [6].

Various molecular 1—3D representations such as SMILES, InChI, or connectivity matrices allow to encode species information into an AI computational framework. From this information, molecular descriptors may be generated and linked to macroscopic properties through quantitative structure-property relationships (QSPRs). This chapter provides an overview of various AI-based methods than can be used to predict fuel properties and thereby enable fuel design.

1.2 Fuel formulation workflow

Data science, which is considered as the fourth pillar of scientific discovery, is generally not part of the training of fuel scientists and dedicated ecosystems fusing computational modeling, virtual high-throughput screening, and big data analytics are still under development [7]. Ideally, fuel formulation would base on an integrated multiscale optimization [8] of fuel properties and production processes to minimize well-to-wheel fuel impact on environment while ensuring economic feasibility (Fig. 3.1). This can be performed through the coupling of computer-aided molecular design and Internal Combustion (IC) engine simulations [13,14] or joint fuel synthetic path—single cylinder combustion optimization [15]. In this coupled workflow, fuel properties are typically evaluated by group contribution (GC) or QSPR models [16] involving molecular descriptors. From validated QSPR models, an inverse-QSPR (i-QSPR) problem can be solved to identify species possessing a targeted property [17]. Advances in computer science have enabled virtual high-throughput screening in which large databases are reduced to a small set of promising molecules for experimentalists to work on [18]. AI developments support this subsequent step through the automation of experimental measurements of fuel properties using robotics [19,20] which may minimize measurement variability [21].

To avoid unbalanced fuel properties at later stages of the computational funnel [22] promising candidates for fuel/engine/aftertreatment (ATS)

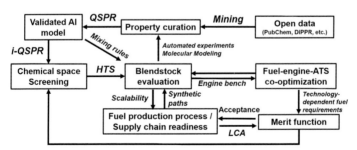

Figure 3.1 Schematic overview of fuel formulation workflow from transportation fuel screening projects. *(Jooß C, Welter F, Leisten I, Richert A, Schaar AK, Valdez AC, et al., Scientific cooperation engineering in the cluster of excellence integrative production*

co-optimization can be selected by introducing a merit function assessing the combined effect of multiple fuel properties p$_i$ on an overall engine-dependent target, which can be engine brake efficiency η_b [11] according to the following expansion:

$$\frac{\Delta \eta_b}{\eta_b} = f(p_1, \ldots, p_n) = \sum_{i=1}^{n} \frac{1}{\eta_b} \frac{\partial \eta_b}{\partial p_i} \left(p_i - p_{i,\text{ref}} \right) \tag{3.1}$$

which highlights a coefficient $\frac{1}{\eta_b} \frac{\partial \eta_b}{\partial p_i}$ for each property. Co-optima project merit function involved RON, octane sensitivity, heat of vaporization, particulate index, and catalyst light-off temperature. Merit function f is then maximized over the chemical space under molecular structure feasibility and properties constraints [23] using QSPR models previously developed for single or multiple fuel properties. It was found that RON exhibits the largest impact on the merit function of all the fuel properties considered, in line with previous findings.

To optimize fuel formulation more holistically, fuel well-to-wheel exergy [24–27]; or environmental impact factor may be used as alternate targets. An adaptive neuro-fuzzy inference systems approach, a combination of neural networks and fuzzy systems, has been recently applied to biofuel life cycle analysis to predict environmental index [27].

1.3 Artificial intelligence modeling approaches

Machine learning (ML) techniques used in fuel formulation involve mostly:
- Ensemble methods such as random forests (RFs)
- Unsupervised techniques such as principal component analysis (PCA) or k-means clustering
- Supervised techniques such as MLR and SVM
- DL techniques such as convolutional neural network (CNN) and recurrent neural network (RNN).
- Reinforcement learning (RL)

---

technology for high-wage countries at RWTH aachen university, Jeschke S, Isenhardt I, Hees F, Henning K, editors. Automation, communication and cybernetics in science and engineering 2013/2014, Springer International Publishing, Cham; 2014. p. 103–109; Hoppe F, Heuser B, Thewes M, Kremer F, Pischinger S, Dahmen M, et al., Tailor-made fuels for future engine concepts. Int J Engine Res 2015;17(1):16–27; Szybist JP, Busch S, McCormick RL, Pihl JA, Splitter DA, Ratcliff MA, et al. What fuel properties enable higher thermal efficiency in spark-ignited engines? Prog Energy Combust Sci 2021;82:100876; Fayet G, Rotureau P. How to use QSPR-type approaches to predict properties in the context of Green Chemistry, Biofuels, Bioprod. Biorefining 2016;10(6):738–752.)

RF mimics human way of thinking by averaging predictions of several decision trees trained on different data samples using a greedy algorithm [28]. This ensemble average (wisdom of crowd effect) is also applied in consensus modeling with sometimes an additional optimization of weighting factors for the different models [29]. K-means clustering method, which groups similar data into evenly sized clusters, is seldom used as a stand-alone ML method for QSPR but is particularly useful during dataset preparation [30,31]. PCA is a method of choice for exploratory analysis as it performs dimension reduction by regressing the input data along the eigenvectors associated with the largest eigenvalues of the covariance matrix. However, it has difficulty accounting for nonlinear relationships between molecular descriptors and chemical and technological properties. Furthermore, it cannot properly consider data that have strong non-orthogonal features or with high variance [32].

MLR assumes a multiple linear relationship between the input variables and dependent output variable. Each of the input variables is multiplied by a slope term with an additional intercept term. The training set is used to determine the weights of the algorithm which minimize the error between the plane corresponding to relationship and the data points. The validation set is used to tune the hyper-parameters such the algorithm structure and regularization parameters. The test set is then used to report the (mean absolute or root mean squared) error of the model on unseen data. Depending on available computational resources, this workflow can be extended to k-fold cross-validation workflow in which the entire dataset is divided into k sets, each set being used as validation and test set once in a loop. MLR fails to capture nonlinear behavior (e.g., azeotrope), but it is advisable to start modeling with MLR before developing a more complex model, unless there is an obvious reason not to do so. Linear models such as MLR are expected to be restricted to the description of the influence of homogeneous descriptors on similar compounds properties over a narrow variation interval in the property space [33].

Fortunately, nonlinearity is readily handled by SVM which designs the best hyperplane separating data into classes by maximizing its distance with the closest points (called support vectors). The performance metrics for SVM classification model are derived from the confusion matrix (true/false positive/negative rates). SVM employs custom kernel functions to map the labeled points of the training set in a higher dimension space where they can be linearly separated. This concept has been applied to kernel partial least squares regression method, which has been implemented industrially [34].

While SVM segregates data in sectors separated by the maximum distance, implying a limited precision in continuous quantities representation, artificial neural networks (ANNs) map continuously input data into outputs. Even if their respective performances are problem dependent [35] and they do not provide any physical intuition, ANNs are less sensitive to noisy data than SVM and allow a straightforward merit function implementation [36]. Due to their ability to adapt to highly nonlinear relations, DL techniques are increasingly used in chemistry [37], although they are expected to require more data than classical ML methods for their training for the same fuel property [38].

A neural network consists of an input layer, several hidden layers of computational nodes and an output layer, corresponding to function composition [39]. Each of the nodes in one layer is interconnected with each and every other node of the next layer as shown in Fig. 3.2. The nodes are created from the original inputs by a weighted sum, and then passed through an activation function f (rectified linear unit, sigmoid, tanh, etc) that allows for nonlinearity in ANN. The training of the network involves finding the summation weights w_{ij} for the nodes in all layers. This is done by backpropagating the loss function gradient with respect to the weights using the chain rule from the fast to the first layer [40]. The tuple of hyperparameters (number of hidden layers, number of nodes in each of the hidden layer, activation function, regularization parameter, learning rate) needs to be carefully chosen while training, either by intuition or dedicated (e.g., Bayesian or evolutionary) optimization algorithm.

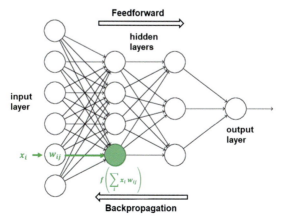

Figure 3.2 Structure of artificial neural network (ANN).

CNNs mimic visual cortex connectivity by involving successively convolutional, pooling, and perceptron layers which respectively filter, reduce feature dimensionality by cluster output combination and classify the image obtained. RNN were developed to model brain processes like memory and attention. RNN inputs have no determined size limit, allowing temporal dynamic behavior. RNN learns typically to predict the probability of the next character in a current SMILES based on previous ones [41]. Particular cases of RNN are long short-term memory (LSTM) networks which have an additional forget gate and gated recurrent units which have a reset and an update gate. Blends may be handled through dedicated permutation-invariant network architecture including mix function to the interact/aggregate blocks and allowing to learn simultaneously on feature description and composition [42].

At the end of the day, the choice of a modeling approach will depend on the amount of data and the expected complexity of the input—output relationship. Usually, simple (but not too simple) models are preferable if the dataset is small. On the other hand, complex models perform generally better on larger datasets at the expense of large computational training cost and more propensity to overfitting, i.e., tendency to predict too closely a particular set of data at hand which results in unreliable prediction of new observations. This can be rationalized by monitoring both training R^2 (which approaches unity with increases model complexity) and validation Q^2 (which typically reaches a maximum with model complexity) squared correlation coefficients.

2. Application of artificial intelligence to fuel formulation

2.1 High throughput screening: finding a needle in the haystack

Chemical space corresponds to the space spanned by any atom combination, estimated to be of the order of 10^{60} for limited weight CHONS compounds [43]. However, ML models are expected to work correctly only within the boundaries of their training dataset, calling for large and more diverse datasets. Experiments or molecular modeling may be performed in the first place to augment this training dataset using a design of experiments method [44,45]. Despite the much lower computational cost of ML models for prediction than training, chemical space enumeration is currently limited to 15—20 heavy atoms due to the combinatorial explosion [46]. Various alternate chemical screening methods were proposed such as

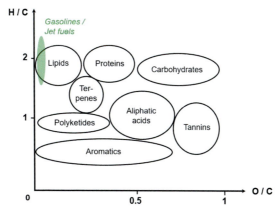

Figure 3.3 Van Krevelen diagram representing schematically the CHO chemical space explored in transportation fuels components screening.

molecular morphing [47] which generates structures connecting the starting and target structure.

To avoid exhaustive enumeration, various mutations/crossover may be applied to initial structure to produce new candidate structures [48], notably through SMARTS formalism, and evaluate the fitness of each molecule in the population [49]. The mutations on a structure are stopped if the said structure exceeds a threshold molecular weight or carbon number, similar to Lipinski and Morowitz rules featured in other fields of chemical research [50]. Fig. 3.3 shows one such threshold on the CHO chemical space explored in transportation fuels components screening on Van Krevelen diagram [51].

Traditional chemical space screening proceeds through topological descriptors and use a molecular graph generator [52]. However, descriptor generation is non-differentiable, preventing inverse mapping from the descriptor space back to molecules [53]. In DL approaches, molecules are typically generated from most probable latent space decoding [54], the differentiability of latent representations allowing gradient-based optimization. While RNN can generate SMILES through decoder, generative adversarial networks (GANs), combining a generator and a discriminator network, can directly generate molecular graphs. RNN-based RL allows to perform genuine QSPR analysis [55], one part of the network generating SMILES and the other predicting properties. Unlike RNN, GAN exempt the user from knowing probability distribution explicitly which is interesting for high-dimensional data [41].

2.2 Fuel property prediction by machine learning models

Table 3.1 presents different approaches attempted to predict fuel properties, highlighting some molecular descriptors common to multiple properties. Most of the studies listed focus on pure species, mixture properties being deduced in turn through a mixing rule at the molecular descriptor or property level [85,92]. Also, few studies focus on simultaneous optimization of multiple fuel end-use properties [93].

However, in absence of detailed composition, fuel properties may be correlated to readily available observables [94]. Mendes et al. [95] used American Society for Testing and Materials (ASTM) distillation curves as input features. Fourier transform infrared [96–100], flame emission [100], nuclear magnetic resonance (NMR) [101–104], dispersive fiber-optic Raman [104], and dielectric spectroscopy [105] as well as GCxGC [106] were processed to generate discriminative input features by different groups. The spectra are either directly used or refined to construct the functional groups which are then used as input features. The later requires a fundamental knowledge on the spectra corresponding to certain functional groups and also selection of important functional groups that affect fuel property. An example of spectra conversion to functional groups is shown Fig. 3.4.

Jameel et al. [103] generated ^{1}H NMR spectra of 128 pure hydrocarbons, 123 hydrocarbon ethanol blends of known composition, and 30 FACE (fuels for advanced combustion engines) gasoline ethanol blends. This was then converted to generate nine structural descriptors (paraffinic primary to tertiary carbons, olefinic, naphthenic, aromatic, and ethanolic OH groups, molecular weight, and branching index). ANN was then used to train RON and MON of the fuels in dataset with the generated structural descriptors. The hyperparameters of ANN that were tuned in this study were the number of units per layer, regularization coefficients, and the number of layers, using the K-fold cross validation. Interestingly, while ANN modeling of RON and MON of these gasolines was successful ($R^2 = 0.99$ for both RON and MON), MLR training over the same dataset with same input features resulted in a poor correlation with $R^2 = 0.51$ for RON and $R^2 = 0.52$ for MON, as shown in Fig. 3.5. This comparison demonstrates the nonlinearity of fuel properties with respect to these structural descriptors requires a good model that can capture the nonlinearity.

Table 3.1 Main approaches attempted to predict selected fuel properties and associated descriptors.

Species property (P_i)	Artificial intelligence approach	Reported descriptors [1]
Octane/cetane number, autoignition metrics	ANN-based group contributions [56], SVM-based on boruta features elimination [57], CNN [58,59], graph NN [60], ANN [61], kNN, RF [62], HDMR/CNN [63,64]	Molecular weight, critical volume, balaban/Kier-Hall/Wiener index, water/octanol partitioning.
Phase equilibria: vapor pressure, boiling point, critical properties	MLR [65], CNN [66], MLR, ANN [67], ANN [68,69], LSTM [70], matrix completion method [71]	Kappa shape index, Wiener and Harary indices.
Rheology: density, viscosity, surface tension	Consensus SVM/ANN prediction [72], ANN metamodel of phenomenological model parameters [73]	Molecular weight, randic index, solvent-accessible surface area
Storage/material compatibility: induction period, swelling	PCA-ANN [74], ANN-SVM-MLR [75]	Excitation-emission matrix, number of sp^3 C, number of C=C bonds
HHV, LHV, HoV	GC [76], SVM, ANN [29,77], ant colony—PLS—MLR [78], MLR [79], GA-SVM [80,81]	Van der waals surface area, number of carbons
Soot index, exhaust aftertreatment activity	Bayesian inference of GC [82], PCA—ANN [83], SVM, RF, PLS [84]	LUMO—HOMO energy, functional groups
Flammability limits, flash point	SVM, ANN [85], SVM [86], ANN—particle swarm [87], kNN, RF, MLR, SVM [88]	Randic index, polarizability, burden matrix, number of sp^3 C
Toxicity	SVM [89], MLR, RNN [90], SVM, RF, gradient boosting, NB, linear discriminant analysis, CNN [91]	Wiener index, LUMO—HOMO energy

ANN, artificial neural network; *SVM*, support vector machine; *NN*, neural network; *CNN*, convolutional neural network; *PLS*, partial least squares; *LSTM*, long short-term memory; *GC*, group contribution; *PCA*, principal component analysis; *RNN*, recurrent neural network

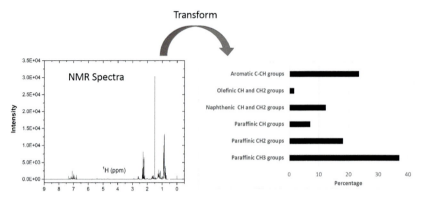

Figure 3.4 A sample of nuclear magnetic resonance spectra and its translation to functional groups.

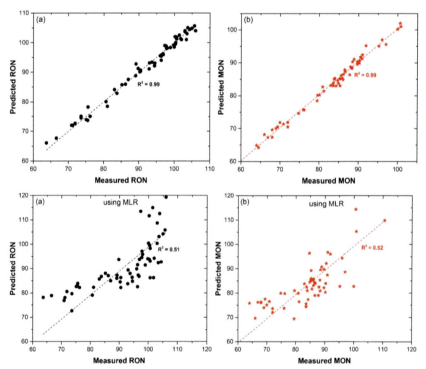

Figure 3.5 Comparison of measured and predicted (A) RON and (B) MON values for artificial neural network (ANN) and multiple linear regression (MLR) models. *(This figure is reproduced from Abdul Jameel AG, Van Oudenhoven V, Emwas AH, Sarathy SM. Predicting octane number using nuclear magnetic resonance spectroscopy and artificial neural networks, Energy Fuels 2018;32(5):6309–6329 with permission from ACS.)*

2.3 Reaction discovery

Designing a new fuel lies not only in finding structures exhibiting required physical property but it consists also in the screening of fuel synthesis, oxidation, and pyrolysis chemical pathways to ensure that reactivity is controlled not only over a few regulatory conditions (e.g., RON, MON, induction delay, FP) but over the whole fuel life cycle and vehicle operation map. This step of the workflow allows an early identification of synergetic effects in fuel combustion and provides adequate combustion thermokinetic models for engine numerical design using computational fluid dynamic (CFD) and fuel property accuracy refinement.

Reaction network generation for fuel combustion or retrosynthesis may be performed by a rule-based expert system exploring a predefined template of reaction families for which the corresponding estimated rate constants may be estimated by on-the-fly quantum electronic structure calculations [107]. The search of optimal retrosynthetic paths may be performed through RL [108]. Transition State geometries may be estimated from GC method [109] or graph neural network [110] prior to Schrödinger equation solution, debottlenecking the $3N_{atom}-6$ dimensional potential energy surface (PES) screening and therefore kinetic rate constant generation. The electronic structure calculation may be made even faster by using a DL model to evaluate directly wavefunction [111]. It is also possible to train neural nets on post-Hartree—Fock PES to achieve chemical accuracy, i.e., error on electronic energy <1 kcal/mol [112]. These models would be also useful during the abovementioned high-throughput screening (HTS) step as it allows to generate quantum descriptors such as frontier orbitals energies at reduced cost [113]. To explore reaction pathways in a physics-agnostic way, quantum or classical molecular dynamics calculations may be performed to find unexpected products from starting molecule without defining reaction coordinate [114,115]. New Neural Network reactive force fields have been shown to be much more accurate than their conventional counterparts [116]. Further, a gaussian process regression of the atomistic PES model would allow a unified treatment for molecules and condensed-phase structures which would pave the way for a smoother exploration of multi-phase engine pollutants and wall deposits evolution [117].

In any case, AI-based methods are not expected to supersede physical methods which seek to explicitly understand the inner workings of a phenomenon, while data-driven methods focus on approximating its outer behavior [118].

2.4 Fuel-engine co-optimization

In a multiscale modeling framework, thermokinetic, transport, and rheological information related to a new fuel derived from AI-enhanced physical models [119] may then be used in reactive CFD models. These fluid dynamics models use increasingly DL to estimate more accurately turbulent subgrid scale fuel combustion rate than their algebraic counterparts [120,121].

Further, the computational burden associated with the stiff chemical kinetic source term integration in Navier—Stokes equations may be largely reduced by either ANN-based HyChem kinetics [122] or direct DL on species profiles [123], resulting in typically two orders of magnitude speedup with respect to stiff chemical kinetics solver. An alternate approach training a CNN on CFD flow field data has been demonstrated recently, albeit on a single fuel [124].

Exhaust aftertreatment system needs also to be optimized given a new (fuel-engine) association. RF models have been successfully trained on heterogeneous kinetic Monte Carlo models for CO catalytic oxidation [125] and implemented in CatalyticFoam CFD code. An ANN model of a selective catalytic reduction on a particulate filter (SCRF) was trained on NEDC simulations using Axisuite aftertreatment simulation software, resulting in accurate predictions with a speedup of 450 [126]. However, the chemical space to screen to optimize catalytic formulation is much larger than for fuels due to a large variety of elements, polymorphs, and crystal defects [127].

3. Conclusions and perspectives

As can be deduced from this short review of an expanding field, fuel discovery and formulation processes are increasing supported by AI-based methods, whether they touch upon data mining, robotics, chemical space exploration, integrated well-to-wheel fuel optimization, etc. While AI provides expanding possibilities to deal with complex physiochemical phenomena involved during fuel life, numerous challenges remain to be met to make AI a daily tool for fuel experts.

First, there is currently no "Swiss army knife" which could address all problems involved in fuel formulation, since datasets for different properties are widely varying in size, homogeneity, and linearity. Fuel experts have been typically juggling between different AI tools and datasets to cover the daunting range of scales, technological readiness levels, and industrial constrains. Academic AI tools are progressively tested in industrial contexts, generating know-how for both communities, bridging the complexity and representativity gap.

Secondly, despite the intense research on fuel description, the current descriptors for mixtures are not fully satisfactory [12]. Similarly in the case of nanomaterials, the development of dedicated descriptors is clearly needed [128]. Yet, the unequivocal definition of mixtures is not a new problem in chemistry [129].

Thirdly, the evolution of the respective role of human researchers and AI in scientific discovery has to be addressed urgently to provide an engaging teleological view of researcher's work. Alan Turing noted that in ML the "teacher will often be very largely ignorant of quite what is going on inside, although he may still be able to some extent to predict his pupil's behavior" [130]. Despite progress in RL [131], human intuition remains necessary to generate intelligibility out of ML model results and steer open-ended research [132]. Machines are certainly freeing up time for humans to solve higher-level questions but at the expense of redefining their connection with respect to the object of their research [133].

A branch in combustion that has not yet adopted AI is robotics. While control systems are largely used in combustion experiments, the experiments are planned and conducted with human interference. AI can be used to optimize the formulations of a certain mixture from the scratch using robotic experiments driven by ML [134]. These methods has been of interest lately for organic synthesis [135,136]. However, combustion experiments are more complex and involve explosive chemicals, making their implementation challenging and lengthy. Another interesting field of AI is active learning, which can be used to plan and design expensive combustion experiments [137]. Data visualization is another branch that has largely evolved due to the advent to CNNs, and this is being used for the determination of several phenomenon in combustion [138,139]. To sum up, AI is being crossed-over into many other fields which are already being used in a diverse combustion field. Therefore, as each of the other fields evolve, the innovations in AI will eventually be disseminated in combustion.

Acknowledgments

The authors thank Nursulu Kuzhagaliyeva and Francesco Tutino for useful discussions.

References

[1] Nieto-Draghi C, Fayet G, Creton B, Rozanska X, Rotureau P, de Hemptinne J-C, et al. A general guidebook for the theoretical prediction of physicochemical properties of chemicals for regulatory purposes. Chem Rev 2015;115(24):13093−164.

[2] George J, Hautier G. Chemist versus machine: traditional knowledge versus machine learning techniques. Trends Chem 2021;3(2):86−95.

[3] Plehiers PP, Symoens SH, Amghizar I, Marin GB, V Stevens C, Van Geem KM. Artificial intelligence in steam cracking modeling: a deep learning algorithm for detailed effluent prediction. Engineering 2019;5(6):1027−40.

[4] Sarathy SM, Farooq A, Kalghatgi GT. Recent progress in gasoline surrogate fuels. Prog Energy Combust Sci 2018;65:1−42.

[5] Ashraf C, Jain A, Xuan Y, van Duin ACT. ReaxFF based molecular dynamics simulations of ignition front propagation in hydrocarbon/oxygen mixtures under high temperature and pressure conditions. Phys Chem Chem Phys 2017;19(7):5004−17.

[6] Han S, Li X, Guo L, Sun H, Zheng M, Ge W. Refining fuel composition of RP-3 chemical surrogate models by reactive molecular dynamics and machine learning. Energy Fuels 2020;34(9):11381−94.

[7] Hachmann J, Afzal MAF, Haghighatlari M, Pal Y. Building and deploying a cyber infrastructure for the data-driven design of chemical systems and the exploration of chemical space. Mol Simulat 2018;44(11):921−9.

[8] König A, Marquardt W, Mitsos A, Viell J, Dahmen M. Integrated design of renewable fuels and their production processes: recent advances and challenges. Curr Opin Chem Eng 2020;27:45−50.

[9] Jooß C, Welter F, Leisten I, Richert A, Schaar AK, Valdez AC, et al. Scientific cooperation engineering in the cluster of excellence integrative production technology for high-wage countries at RWTH aachen university. In: Jeschke S, Isenhardt I, Hees F, Henning K, editors. Automation, communication and cybernetics in science and engineering 2013/2014. Cham: Springer International Publishing; 2014. p. 103−9.

[10] Hoppe F, Heuser B, Thewes M, Kremer F, Pischinger S, Dahmen M, et al. Tailor-made fuels for future engine concepts. Int J Engine Res 2015;17(1):16−27.

[11] Szybist JP, Busch S, McCormick RL, Pihl JA, Splitter DA, Ratcliff MA, et al. What fuel properties enable higher thermal efficiency in spark-ignited engines? Prog Energy Combust Sci 2021;82:100876.

[12] Fayet G, Rotureau P. How to use QSPR-type approaches to predict properties in the context of Green Chemistry, Biofuels, Bioprod. Biorefining 2016;10(6):738−52.

[13] Hada S, Solvason CC, Eden MR. Characterization-based molecular design of biofuel additives using chemometric and property clustering techniques. Front Energy Res 2014;2:20.

[14] Gschwend D. A systematic search for next generation transportation fuels. PhD Thesis. ETH Zurich; 2018.

[15] Heuser B, Kremer F, Pischinger S, Julis J, Leitner W. Optimization of diesel combustion and emissions with newly derived biogenic alcohols. SAE/KSAE 2013 International Powertrains, Fuels & Lubricants Meeting. SAE International; 2013.

[16] Dahmen M, Marquardt W. Model-based formulation of biofuel blends by simultaneous product and pathway design. Energy Fuels 2017;31(4):4096−121.

[17] Gantzer P, Creton B, Nieto-Draghi C. Inverse-QSPR for de novo design: a Review. Mol Inform 2020;39(4):1900087.
[18] Pyzer-Knapp EO, Suh C, Gómez-Bombarelli R, Aguilera-Iparraguirre J, Aspuru-Guzik A. What is high-throughput virtual screening? A perspective from organic materials discovery. Annu Rev Mater Res 2015;45(1):195−216.
[19] Burger B, Maffettone PM, V Gusev V, Aitchison CM, Bai Y, Wang X, et al. A mobile robotic chemist. Nature 2020;583(7815):237−41.
[20] Delhaye D, Ouf F-X, Ferry D, Ortega IK, Penanhoat O, Peillon S, et al. The MERMOSE project: characterization of particulate matter emissions of a commercial aircraft engine. J Aerosol Sci 2017;105:48−63.
[21] Gromski PS, Henson AB, Granda JM, Cronin L. How to explore chemical space using algorithms and automation. Nat Rev Chem 2019;3(2):119−28.
[22] Bleicher KH, Böhm H-J, Müller K, Alanine AI. Hit and lead generation: beyond high-throughput screening. Nat Rev Drug Discov 2003;2(5):369−78.
[23] Gertig C, Fleitmann L, Hemprich C, Hense J, Bardow A, Leonhard K. Integrated in silico design of catalysts and processes based on quantum chemistry. Comp Aid Chem Engi. 2020;48:889−94.
[24] Flórez-Orrego D, Silva JAM, de Oliveira Jr S. Exergy and environmental comparison of the end use of vehicle fuels: the Brazilian case. Energy Convers Manag 2015;100:220−31.
[25] de Oliveira Junior S. Exergy: production, cost and renewability. Springer Science & Business Media; 2012.
[26] Portha J-F, Louret S, Pons M-N, Jaubert J-N. Estimation of the environmental impact of a petrochemical process using coupled LCA and exergy analysis. Resour Conserv Recycl 2010;54(5):291−8.
[27] Mayol AP, Juan JLGS, Sybingco E, Bandala A, Dadios E, Ubando AT, et al. Environmental impact prediction of microalgae to biofuels chains using artificial intelligence: a life cycle perspective. IOP Conf Ser Earth Environ Sci 2020;463:12011.
[28] Breiman L. Random forests. Mach Learn 2001;45(1):5−32.
[29] Saldana DA, Starck L, Mougin P, Rousseau B, Creton B. On the rational formulation of alternative fuels: melting point and net heat of combustion predictions for fuel compounds using machine learning methods. SAR QSAR Environ Res 2013;24(4):259−77.
[30] Leonard JT, Roy K. On selection of training and test sets for the development of predictive QSAR models. QSAR Comb Sci 2006;25(3):235−51.
[31] Yuan S, Zhang Z, Sun Y, Kwon JS-I, V Mashuga C. Liquid flammability ratings predicted by machine learning considering aerosolization. J Hazard Mater 2020;386:121640.
[32] Jolliffe IT. A note on the use of principal components in regression. J R Stat Soc Ser C (Appl Stat) 1982;31(3):300−3.
[33] Saldana DA, Starck L, Mougin P, Rousseau B, Pidol L, Jeuland N, et al. Flash point and cetane number predictions for fuel compounds using quantitative structure property relationship (QSPR) methods. Energy Fuels 2011;25(9):3900−8.
[34] An Y, Sherman W, Dixon SL. Kernel-based partial least squares: application to fingerprint-based QSAR with model visualization. J Chem Inf Model 2013;53(9):2312−21.
[35] Shao Y, Lunetta RS. Comparison of support vector machine, neural network, and CART algorithms for the land-cover classification using limited training data points. ISPRS J Photogram. Remote Sens 2012;70:78−87.
[36] Goodfellow I, Bengio Y, Courville A. Deep learning. MIT Press; 2016.
[37] Hoskins JC, Himmelblau DM. Artificial neural network models of knowledge representation in chemical engineering. Comput Chem Eng 1988;12(9):881−90.

[38] Roh Y, Heo G, Whang SE. A survey on data collection for machine learning: a big data - AI integration perspective. IEEE Trans Knowl Data Eng 2019;33(4):1328−47.
[39] Barradas Filho AO, Barros AKD, Labidi S, Viegas IMA, Marques DB, Romariz ARS, et al. Application of artificial neural networks to predict viscosity, iodine value and induction period of biodiesel focused on the study of oxidative stability. Fuel 2015;145:127−35.
[40] Lecun Y, Bottou L, Bengio Y, Haffner P. Gradient-based learning applied to document recognition. Proc IEEE 1998;86(11):2278−324.
[41] Sattarov B, Baskin II, Horvath D, Marcou G, Bjerrum EJ, Varnek A. De novo molecular design by combining deep autoencoder recurrent neural networks with generative topographic mapping. J Chem Inf Model 2019;59(3):1182−96.
[42] Hanaoka K. Deep neural networks for multicomponent molecular systems. ACS Omega 2020;5(33):21042−53.
[43] Bohacek RS, McMartin C, Guida WC. The art and practice of structure-based drug design: a molecular modeling perspective. Med Res Rev 1996;16(1):3−50.
[44] Davis R, John P. Application of Taguchi-based design of experiments for industrial chemical processes. Stat Approach Emphas Des Exp Appl Chem Process 2018;137:137−55.
[45] Myers RH, Montgomery DC. Response surface methodology: process and product in optimization using designed experiments. 1st ed. USA: John Wiley & Sons, Inc.; 1995.
[46] van Deursen R, Reymond J-L. Chemical space travel. ChemMedChem 2007;2(5):636−40.
[47] Hoksza D, Svozil D. Exploration of chemical space by molecular morphing. In: 2011 IEEE 11th international conference on bioinformatics and bioengineering; 2011. p. 201−8.
[48] Virshup AM, Contreras-García J, Wipf P, Yang W, Beratan DN. Stochastic voyages into uncharted chemical space produce a representative library of all possible drug-like compounds. J Am Chem Soc 2013;135(19):7296−303.
[49] Cheng CY, Campbell JE, Day GM. Evolutionary chemical space exploration for functional materials: computational organic semiconductor discovery. Chem Sci 2020;11(19):4922−33.
[50] Morowitz HJ, Kostelnik JD, Yang J, Cody GD. The origin of intermediary metabolism. Proc Natl Acad Sci Unit States Am 2000;97(14):7704−8.
[51] Burnham AK. In: Sorkhabi R, editor. Van Krevelen diagrams BT - encyclopedia of petroleum geoscience. Cham: Springer International Publishing; 2018. p. 1−5.
[52] Churchwell CJ, Rintoul MD, Martin S, Visco DP, Kotu A, Larson RS, et al. The signature molecular descriptor: 3. Inverse-quantitative structure−activity relationship of ICAM-1 inhibitory peptides. J Mol Graph Model 2004;22(4):263−73.
[53] Popova M, Isayev O, Tropsha A. Deep reinforcement learning for de novo drug design. Sci Adv 2018;4(7).
[54] Gómez-Bombarelli R, Wei JN, Duvenaud D, Hernández-Lobato JM, Sánchez-Lengeling B, Sheberla D, et al. Automatic chemical design using a data-driven continuous representation of molecules. ACS Cent Sci 2018;4(2):268−76.
[55] Olivecrona M, Blaschke T, Engkvist O, Chen H. Molecular de-novo design through deep reinforcement learning. J Cheminf 2017;9(1):48.
[56] Kubic WL, Jenkins RW, Moore CM, Semelsberger TA, Sutton AD. Artificial neural network based group contribution method for estimating cetane and octane numbers of hydrocarbons and oxygenated organic compounds. Ind Eng Chem Res 2017;56(42):12236−45.
[57] Liu Z, Zhang L, Elkamel A, Liang D, Zhao S, Xu C, et al. Multiobjective feature selection approach to quantitative structure property relationship models for predicting the octane number of compounds found in gasoline. Energy Fuel 2017;31(6):5828−39.

[58] Buras ZJ, Safta C, Zádor J, Sheps L. Simulated production of OH, HO$_2$, CH$_2$O, and CO$_2$ during dilute fuel oxidation can predict 1st-stage ignition delays. Combust Flame 2020;216:472—84.
[59] vom Lehn F, Brosius B, Broda R, Cai L, Pitsch H. Using machine learning with target-specific feature sets for structure-property relationship modeling of octane numbers and octane sensitivity. Fuel 2020;281:118772.
[60] Schweidtmann AM, Rittig JG, König A, Grohe M, Mitsos A, Dahmen M. Graph neural networks for prediction of fuel ignition quality. Energy Fuel 2020;34(9):11395—407.
[61] Kessler T, Sacia ER, Bell AT, Mack JH. Artificial neural network based predictions of cetane number for furanic biofuel additives. Fuel 2017;206:171—9.
[62] Whitmore LS, Davis RW, McCormick RL, Gladden JM, Simmons BA, George A, et al. BioCompoundML: a general biofuel property screening tool for biological molecules using random forest classifiers. Energy Fuel 2016;30(10):8410—8.
[63] Han W, Sun Z, Scholtissek A, Hasse C. Machine Learning of ignition delay times under dual-fuel engine conditions. Fuel 2021;288:119650.
[64] Albahri TA. Structural group contribution method for predicting the octane number of pure hydrocarbon liquids. Ind Eng Chem Res 2003;42(3):657—62.
[65] Katritzky AR, Slavov SH, Dobchev DA, Karelson M. Rapid QSPR model development technique for prediction of vapor pressure of organic compounds. Comput Chem Eng 2007;31(9):1123—30.
[66] McClelland HE, Jurs PC. Quantitative structure—property relationships for the prediction of vapor pressures of organic compounds from molecular structures. J Chem Inf Comput Sci 2000;40(4):967—75.
[67] Gao J, Wang X, Yu X, Li X, Wang H. Calculation of polyamides melting point by quantum-chemical method and BP artificial neural networks. J Mol Model 2006;12(4):521—7.
[68] Hall LH, Story CT. Boiling point and critical temperature of a heterogeneous data set: QSAR with atom type electrotopological state indices using artificial neural networks. J Chem Inf Comput Sci 1996;36(5):1004—14.
[69] Li G, Hu Z, Hou F, Li X, Wang L, Zhang X. Machine learning enabled high-throughput screening of hydrocarbon molecules for the design of next generation fuels. Fuel 2020;265:116968.
[70] Su Y, Wang Z, Jin S, Shen W, Ren J, Eden MR. An architecture of deep learning in QSPR modeling for the prediction of critical properties using molecular signatures. AIChE J 2019;65(9):e16678.
[71] Jirasek F, Alves RAS, Damay J, Vandermeulen RA, Bamler R, Bortz M, et al. Machine learning in thermodynamics: prediction of activity coefficients by matrix completion. J Phys Chem Lett 2020;11(3):981—5.
[72] Saldana DA, Starck L, Mougin P, Rousseau B, Ferrando N, Creton B. Prediction of density and viscosity of biofuel compounds using machine learning methods. Energy Fuels 2012;26(4):2416—26.
[73] Cai G, Liu Z, Zhang L, Zhao S, Xu C. Quantitative structure—property relationship model for hydrocarbon liquid viscosity prediction. Energy Fuels 2018;32(3):3290—8.
[74] Kosir S, Heyne J, Graham J. A machine learning framework for drop-in volume swell characteristics of sustainable aviation fuel. Fuel 2020;274:117832.
[75] Sanaeifar A, Jafari A. Determination of the oxidative stability of olive oil using an integrated system based on dielectric spectroscopy and computer vision. Inf Process Agric 2019;6(1):20—5.
[76] Frutiger J, Marcarie C, Abildskov J, Sin G. A comprehensive methodology for development, parameter estimation, and uncertainty analysis of group contribution based property models—an application to the heat of combustion. J Chem Eng Data 2016;61(1):602—13.

[77] Yalamanchi KK, van Oudenhoven VCO, Tutino F, Monge-Palacios M, Alshehri A, Gao X, et al. Machine learning to predict standard enthalpy of formation of hydrocarbons. J Phys Chem A 2019;123(38):8305−13.

[78] Pan Y, Jiang JC, Wang R, Jiang JJ. Predicting the net heat of combustion of organic compounds from molecular structures based on ant colony optimization. J Loss Prev Process Ind 2011;24(1):85−9.

[79] Sosnowska A, Barycki M, Jagiello K, Haranczyk M, Gajewicz A, Kawai T, et al. Predicting enthalpy of vaporization for persistent organic pollutants with quantitative structure−property relationship (QSPR) incorporating the influence of temperature on volatility. Atmos Environ 2014;87:10−8.

[80] Yalamanchi KK, Monge-Palacios M, van Oudenhoven VCO, Gao X, Sarathy SM. Data science approach to estimate enthalpy of formation of cyclic hydrocarbons. J Phys Chem A 2020;124(31):6270−6.

[81] Aldosari MN, Yalamanchi KK, Gao X, Sarathy SM. Predicting entropy and heat capacity of hydrocarbons using machine learning. Energy AI 2021;4:100054.

[82] Das DD, John PCS, McEnally CS, Kim S, Pfefferle LD. Measuring and predicting sooting tendencies of oxygenates, alkanes, alkenes, cycloalkanes, and aromatics on a unified scale. Combust Flame 2018;190:349−64.

[83] Smith A, Keane A, Dumesic JA, Huber GW, Zavala VM. A machine learning framework for the analysis and prediction of catalytic activity from experimental data. Appl Catal B Environ 2020;263:118257.

[84] Takigawa I, Shimizu K, Tsuda K, Takakusagi S. Machine learning predictions of factors affecting the activity of heterogeneous metal catalysts. In: Tanaka I, editor. Nanoinformatics. Singapore: Springer Singapore; 2018. p. 45−64.

[85] Saldana DA, Starck L, Mougin P, Rousseau B, Creton B. Prediction of flash points for fuel mixtures using machine learning and a novel equation. Energy Fuels 2013;27(7):3811−20.

[86] Pan Y, Jiang J, Wang R, Cao H, Zhao J. Quantitative structure−property relationship studies for predicting flash points of organic compounds using support vector machines. QSAR Comb Sci 2008;27(8):1013−9.

[87] Lazzús JA. Prediction of flammability limit temperatures from molecular structures using a neural network−particle swarm algorithm. J Taiwan Inst Chem Eng 2011;42(3):447−53.

[88] Yuan S, Jiao Z, Quddus N, Kwon JS-I, V Mashuga C. Developing quantitative structure−property relationship models to predict the upper flammability limit using machine learning. Ind Eng Chem Res 2019;58(8):3531−7.

[89] Carlsson L, Helgee EA, Boyer S. Interpretation of nonlinear QSAR models applied to ames mutagenicity data. J Chem Inf Model 2009;49(11):2551−8.

[90] Bertinetto CG. Prediction of the physico-chemical properties of low and high molecular weight compounds. Lambert Academic Publishing, 2010.

[91] Jiao Z, Hu P, Xu H, Wang Q. Machine learning and deep learning in chemical health and safety: a systematic review of techniques and applications. ACS Chem Heal Saf 2020;27(6):316−34.

[92] Muratov EN, V Varlamova E, Artemenko AG, Polishchuk PG, Kuz'min VE. Existing and developing approaches for QSAR analysis of mixtures. Mol Inform 2012;31(3-4):202−21.

[93] Cai G, Liu Z, Zhang L, Shi Q, Zhao S, Xu C. Systematic performance evaluation of gasoline molecules based on quantitative structure-property relationship models. Chem Eng Sci 2021;229:116077.

[94] Al-Fahemi JH, Albis NA, Gad EAM. QSPR models for octane number prediction. In: Sajan D, editor. J Theor Chem, vol. 2014; 2014. p. 520652.

[95] Mendes G, Aleme HG, Barbeira PJS. Determination of octane numbers in gasoline by distillation curves and partial least squares regression. Fuel 2012;97:131—6.
[96] Andrade JM, Muniategui S, Prada D. Prediction of clean octane numbers of catalytic reformed naphthas using FT-m.i.r. and PLS. Fuel 1997;76(11):1035—42.
[97] Daly SR, Niemeyer KE, Cannella WJ, Hagen CL. Predicting fuel research octane number using Fourier-transform infrared absorption spectra of neat hydrocarbons. Fuel 2016;183:359—65.
[98] Palani R, AbdulGani A, Balasubramanian N. Treatment of tannery effluent using a rotating disc electrochemical reactor. Water Environ Res 2017;89(1):77—85.
[99] Abdul Jameel AG, Han Y, Brignoli O, Telalović S, Elbaz AM, Im HG, et al. Heavy fuel oil pyrolysis and combustion: kinetics and evolved gases investigated by TGA-FTIR. J Anal Appl Pyrolysis 2017;127:183—95.
[100] de Paulo JM, Barros JEM, Barbeira PJS. A PLS regression model using flame spectroscopy emission for determination of octane numbers in gasoline. Fuel 2016;176:216—21.
[101] Abdul Jameel AG, Elbaz AM, Emwas A-H, Roberts WL, Sarathy SM. Calculation of average molecular parameters, functional groups, and a surrogate molecule for heavy fuel oils using 1H and 13C nuclear magnetic resonance spectroscopy. Energy Fuels 2016;30(5):3894—905.
[102] Abdul Jameel AG, Sarathy SM. Prediction of RON and Mon of gasoline-ethanol using 1 H NMR spectroscopy. Proc Eur Combust Meet 2017:PS1—07—12.
[103] Abdul Jameel AG, Van Oudenhoven V, Emwas A-H, Sarathy SM. Predicting octane number using nuclear magnetic resonance spectroscopy and artificial neural networks. Energy Fuels 2018;32(5):6309—29.
[104] Flecher PE, Welch WT, Albin S, Cooper JB. Determination of octane numbers and Reid vapor pressure in commercial gasoline using dispersive fiber-optic Raman spectroscopy. Spectrochim Acta Part A Mol Biomol Spectrosc 1997;53(2):199—206.
[105] Guan L, Feng XL, Li ZC, Lin GM. Determination of octane numbers for clean gasoline using dielectric spectroscopy. Fuel 2009;88(8):1453—9.
[106] Shi X, Li H, Song Z, Zhang X, Liu G. Quantitative composition-property relationship of aviation hydrocarbon fuel based on comprehensive two-dimensional gas chromatography with mass spectrometry and flame ionization detector. Fuel 2017;200:395—406.
[107] Van de Vijver R, Zádor J, KinBot. Automated stationary point search on potential energy surfaces. Comput Phys Commun 2020;248:106947.
[108] Schreck JS, Coley CW, Bishop KJM. Learning retrosynthetic planning through simulated experience. ACS Cent Sci 2019;5(6):970—81.
[109] Bhoorasingh PL, West RH. Transition state geometry prediction using molecular group contributions. Phys Chem Chem Phys 2015;17(48):32173—82.
[110] Pattanaik L, Ingraham JB, Grambow CA, Green WH. Generating transition states of isomerization reactions with deep learning. Phys Chem Chem Phys 2020;22(41):23618—26.
[111] Schütt KT, Gastegger M, Tkatchenko A, Müller K-R, Maurer RJ. Unifying machine learning and quantum chemistry with a deep neural network for molecular wavefunctions. Nat Commun 2019;10(1):5024.
[112] Smith JS, Nebgen BT, Zubatyuk R, Lubbers N, Devereux C, Barros K, et al. Approaching coupled cluster accuracy with a general-purpose neural network potential through transfer learning. Nat Commun 2019;10(1):2903.
[113] Hansen K, Biegler F, Ramakrishnan R, Pronobis W, von Lilienfeld OA, Müller K-R, et al. Machine learning predictions of molecular properties: accurate many-body potentials and nonlocality in chemical space. J Phys Chem Lett 2015;6(12):2326—31.

[114] Wang L-P, Titov A, McGibbon R, Liu F, Pande VS, Martínez TJ. Discovering chemistry with an ab initio nanoreactor. Nat Chem 2014;6(12):1044–8.
[115] Döntgen M, Przybylski-Freund M-D, Kröger LC, Kopp WA, Ismail AE, Leonhard K. Automated discovery of reaction pathways, rate constants, and transition states using reactive molecular dynamics simulations. J Chem Theor Comput 2015;11(6):2517–24.
[116] Yoo P, Sakano M, Desai S, Islam MM, Liao P, Strachan A. Neural network reactive force field for C, H, N, and O systems. npj Comput Mater 2021;7(1):9.
[117] Bartók AP, De S, Poelking C, Bernstein N, Kermode JR, Csányi G, et al. Machine learning unifies the modeling of materials and molecules. Sci Adv 2017;3(12):e1701816.
[118] Toyao T, Maeno Z, Takakusagi S, Kamachi T, Takigawa I, Shimizu K. Machine learning for catalysis informatics: recent applications and prospects. ACS Catal 2020;10(3):2260–97.
[119] Reynel-Ávila HE, Bonilla-Petriciolet A, Tapia-Picazo JC. An artificial neural network-based NRTL model for simulating liquid-liquid equilibria of systems present in biofuels production. Fluid Phase Equil 2019;483:153–64.
[120] Lapeyre CJ, Misdariis A, Cazard N, Veynante D, Poinsot T. Training convolutional neural networks to estimate turbulent sub-grid scale reaction rates. Combust Flame 2019;203:255–64.
[121] Pulga L, Bianchi GM, Falfari S, Forte C. A machine learning methodology for improving the accuracy of laminar flame simulations with reduced chemical kinetics mechanisms. Combust Flame 2020;216:72–81.
[122] Ranade R, Alqahtani S, Farooq A, Echekki T. An extended hybrid chemistry framework for complex hydrocarbon fuels. Fuel 2019;251:276–84.
[123] Sharma AJ, Johnson RF, Kessler DA, Moses A. Deep learning for scalable chemical kinetics, AIAA scitech 2020 forum. American Institute of Aeronautics and Astro-nauticsIn; 2020 (AIAA SciTech Forum).
[124] An J, Wang H, Liu B, Luo KH, Qin F, He GQ. A deep learning framework for hydrogen-fueled turbulent combustion simulation. Int J Hydrogen Energy 2020;45(35):17992–8000.
[125] Bracconi M, Maestri M. Training set design for machine learning techniques applied to the approximation of computationally intensive first-principles kinetic models. Chem Eng J 2020;400:125469.
[126] März C, Werfel J, Kühne J, Scholz R. Approaches for a new generation of fast-computing catalyst models. Emiss Control Sci Technol 2020;6(2):254–68.
[127] Pfund LY, Matzger AJ. Towards exhaustive and automated high-throughput screening for crystalline polymorphs. ACS Comb Sci 2014;16(7):309–13.
[128] Toropova AP, Toropov AA. QSPR and nano-QSPR: what is the difference? J Mol Struct 2019;1182:141–9.
[129] Needham P. Is water a mixture? Bridging the distinction between physical and chemical properties. Stud Hist Philos Sci 2008;39(1):66–77.
[130] Turing AM. Computing machinery and intelligence. Mind 1950;49(236):433–60. LIX.
[131] Shanahan M, Crosby M, Beyret B, Cheke L. Artificial intelligence and the common sense of animals. Trends Cognit Sci 2020;24(11):862–72.
[132] Coley CW, Eyke NS, Jensen KF. Autonomous discovery in the chemical sciences Part II: outlook. Angew Chem Int Ed 2020;59(52):23414–36.
[133] Lowe D. AI designs organic syntheses. Nature 2018;555(7698):592–3.
[134] Cao L, Russo D, Felton K, Salley D, Sharma A, Keenan G, et al. Optimization of formulations using robotic experiments driven by machine learning DoE. Cell Reports Phys Sci 2021;2(1):100295.

[135] Granda JM, Donina L, Dragone V, Long D-L, Cronin L. Controlling an organic synthesis robot with machine learning to search for new reactivity. Nature 2018;559(7714):377—81.
[136] Clayton AD, Manson JA, Taylor CJ, Chamberlain TW, Taylor BA, Clemens G, et al. Algorithms for the self-optimisation of chemical reactions. React Chem Eng 2019;4(9):1545—54.
[137] Pantazi XE, Moshou D, Kateris D, Gravalos I, Xyradakis P. Automatic identification of gasoline — biofuel blend type in an internal combustion four-stroke engine based on unsupervised novelty detection and active learning. Procedia Technol 2013;8:229—37.
[138] Han Z, Li J, Zhang B, Hossain MM, Xu C. Prediction of combustion state through a semi-supervised learning model and flame imaging. Fuel 2021;289:119745.
[139] Hanuschkin A, Zündorf S, Schmidt M, Welch C, Schorr J, Peters S, et al. Investigation of cycle-to-cycle variations in a spark-ignition engine based on a machine learning analysis of the early flame kernel. Proc Combust Inst 2020;38(4):5751—9.

SECTION 2

Artificial Intelligence and computational fluid dynamics to optimize internal combustion engines

CHAPTER 4

Engine optimization using computational fluid dynamics and genetic algorithms

Alberto Broatch[1], Ricardo Novella[1], José M. Pastor[1], Josep Gomez-Soriano[1] and Peter Kelly Senecal[2]

[1]CMT—Thermal Motors, Polytechnic University of Valencia, Camino de Vera, Valencia, Spain; [2]Convergent Science, Inc., Madison, WI, United States

1. Introduction

The internal combustion engine (ICE) has undergone tremendous advancement since its invention more than a century ago. From how the air is delivered to how the fuel is introduced to the design of the combustion chamber, nearly every aspect has evolved to form a far cleaner and more efficient machine. Historically, much of the advancement was accomplished through experimental trial and error, gut feeling, or sheer perseverance. However, with the invention and wide adoption of computers, a new tool has emerged to help improve engines: computational fluid dynamics (CFD). Engines are, at their core, thermofluid systems, and their behavior is governed by the conservation of mass, momentum, and energy. These equations are far too complex to solve analytically, and as a result, a mesh (i.e., a discretized version of the engine geometry) must be generated so that numerical methods can be applied. Doing so provides a wealth of information far beyond what can be measured experimentally. The velocity field, turbulence quantities, and chemical species are tracked at every point in space and time (to within the mesh resolution and time-steps used). These data allow us to not only see *how* engine performance responds to certain design changes but also provides a deeper understanding of *why*. This level of insight enables us to push engines further than otherwise possible, giving us a much-needed boost in our race toward cleaner and greener propulsion systems.

From a big-picture perspective, engines are relatively simple machines. Air and fuel are brought into the cylinder; the mixture is compressed, ignited, and combusted; and finally, the exhaust gases are released. The technology to make all of this happen quickly and efficiently, however, is far from simple—there are numerous design parameters that can be modified or tuned to improve performance. Many of the parameters (so-called "calibration parameters") can be changed easily on a real engine in a laboratory to assess performance; however, this method does not provide the detailed information necessary to understand why a change in performance is seen. In addition, parameters related to hardware design (piston bowl shape, port geometry, spark plug location, injector location/number of holes …) can be difficult and expensive to test experimentally.

This is where computer modeling—in this case CFD—comes in. CFD allows all of this to be done virtually. It is far easier (at least in theory) to modify a CAD geometry than it is to cut hardware. Of course, it's still not *easy* even with CFD, but modern software has made it significantly more practical. Advancements such as autonomous meshing, adaptive mesh refinement (AMR), and high-performance computing have opened up new realms of possibility. While simulating an engine cycle will never (at least not in the foreseeable future) be as fast as running an engine on the testbench, the total time to reach a final design can be reduced significantly with CFD.

Obviously, CFD predictions are only as good as the input parameters and the software used. Coarse grids, inaccurate models, and overly diffusive numerics can pollute the quality of a simulation. Therefore, a well-validated baseline simulation should be the foundation of any CFD design study. This baseline should include proven models, knowledge of grid convergence, and an assessment of the accuracy/runtime tradeoff (note that accepting some inaccuracy in favor of a faster runtime may not be a bad thing as long as the implications are well understood). Modeling approaches such as finer meshes, larger chemical mechanisms, and large-eddy simulations (as opposed to RANS) promote accuracy, but have been under-used in the past because they require long runtimes. Fortunately, with constantly improving computer hardware, faster solvers, and massive scalability, we are now able to run highly accurate simulations in reasonable wall-clock times.

But how do we harness the power of CFD to design an engine? Trial and error is still sometimes used, but increasingly optimization techniques are used to help guide a design.

Generally, these optimization techniques can be classified as non-evolutionary or evolutionary. In nonevolutionary methods, the set of input parameter combinations to be evaluated are often predefined, before running the simulations. On the other hand, with evolutionary methods the ranges of input parameters are predefined, but the combinations to be evaluated are set in parallel to the simulations. The key difference between these approaches consists of the possibility of integrating real-time information to drive the optimization which is provided by the evolutionary methods.

The classical design of experiments (DoEs) [45,46] combined with the response surface method (RSM) is perhaps the best known nonevolutionary optimization technique. This approach has been applied to a wide range of problems due to its robust statistical foundation and its relative simplicity. In this method, the combinations of inputs to be evaluated are defined according to a given pattern, known as a "design," that provides desirable mathematical and statistical properties. The most usual designs are the central composite design and the Box—Behnken design. In this approach, only a very limited combination of inputs is directly simulated, and thus the probability of finding the optimum set of inputs is extremely low. The RSM provides an additional fitting model based on polynomial regressions relating the target output with the inputs. The search for the optimal combination of inputs is then performed over the fitting models.

Two main aspects should be considered in the application of the DoE method. First, it requires a minimum level of knowledge about the characteristics of the system under optimization. This assures a suitable definition of the fixed combinations of inputs (to avoid unrealistic designs, for example) and confirms that the relation between the target outputs and the inputs corresponds to a smooth (usually second order) polynomial. Second, the number of simulations is determined and increases exponentially with the number of inputs, while most of the simulations at the end are far from the optimum region.

Given these considerations, the DoE method is more suitable for refining a well-known problem, considering a limited number of inputs with small ranges of variation. An example might be the updating of the combustion system of a compression ignition (CI) engine operating with the conventional diesel combustion concept. However, many modern computational engine optimization problems are not suitable for the direct application of DoE techniques as they do not fulfill the previous conditions. An interesting example might be the optimization of an engine operating with an advanced combustion concept [e.g., gasoline compression ignition (GCI), reactivity controlled compression ignition (RCCI)], which is relatively unknown and very sensitive to a great number of inputs. In this context, evolutionary methods such as genetic algorithms (GAs, the focus of this chapter) or particle swarm optimization offer much more flexibility as described later in this chapter.

2. Modeling framework and acceleration strategies

Before an evolutionary method can be applied to engine optimization, a robust, accurate, and efficient modeling framework must be developed. For our purposes, this includes a set of acceleration techniques for the CFD calculation and an appropriate way to parameterize the inputs of interest.

In this section, we discuss a few of the more relevant acceleration strategies that allow CFD to be used in a practical optimization study. Also, we will discuss some of the parameterization schemes and supplementary tools that have been used to complement CFD for engine optimization.

2.1 Computational fluid dynamics acceleration techniques
2.1.1 Adaptive mesh refinement

For CFD to provide meaningful results for engine optimization and design, simulations must be accurate *and* efficient. Historically, these two requirements have been at odds with each other—more accuracy requires higher mesh resolution which results in longer runtimes.

One of the most powerful tools for addressing the accuracy/runtime tradeoff is AMR. The concept of AMR is not new. For example, the book *Structured Adaptive Mesh Refinement (SAMR) Grid Methods*, edited by Baden et al., is based on the proceedings of an Institute for Mathematics and its Applications workshop held in 1997 [1]. About 10 papers covering the programming complexities and numerical challenges of AMR are included, representing the state-of-the-art at the time.

Prior to AMR, an engine mesh was generated by placing resolution based on prior experience or quite simply by guessing. Making a mesh was difficult, so grid convergence was rarely checked. This made it challenging to ascertain the accuracy of any simulation. With the advent of autonomous meshing and AMR, accurate, grid-converged solutions could be achieved with reasonable runtimes.

Put simply, AMR locally refines the mesh in areas with complex flow physics without unnecessarily slowing the simulation with a globally refined grid. A complete AMR algorithm also coarsens areas that are no longer interesting to conserve total cell count.

The use of AMR in engine simulations with complex geometries and moving boundaries was pioneered by Senecal et al. [2] and Richards et al. [3] in the commercial CFD solver, CONVERGE. In CONVERGE, AMR is used to add grid resolution where the flow field is most under-resolved. This is established by calculating the subgrid field throughout the domain and determining where it is the largest. Large subgrid fields correspond to where the curvature of a specified field variable is the highest.

An example of AMR applied to an engine simulation is shown in Fig. 4.1. Here the AMR has been applied to the temperature field and adds resolution at the flame front.

2.1.2 Detailed chemistry acceleration strategies

Just as more mesh improves the accuracy of engine flow field and flame front predictions, more detailed chemistry improves the accuracy of the combustion and emissions calculations. Unfortunately, adding detailed chemistry can significantly add to computation times. However, several techniques exist to speed up chemistry calculations in CFD, allowing larger mechanisms to be used in optimization studies. These techniques include (but are not limited to):

(1) Analytical Jacobian, used for solving the nonlinear system of equations [3].
(2) Sparse iterative linear solver, using a Krylov solver [4], for solving the linear system of a large kinetic mechanisms (>100 species).
(3) Advanced preconditioner, specialized to significantly reduce the computational time when solving extremely large chemical reaction mechanisms (>500 species).

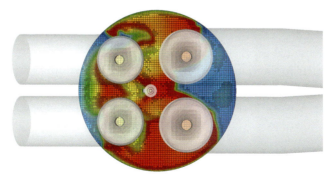

Figure 4.1 Example of a spark-ignited engine simulation with adaptive mesh refinement (AMR). In this case, the colors represent the temperature field (*blue*: low, *red*: high). Enhanced mesh resolution can be observed at the flame front, as determined by the CONVERGE AMR algorithm.

(4) Adaptive zoning [5] accelerates the chemistry calculations by grouping together similar computational cells and then invoking the chemistry solver once per group rather than once per cell.

(5) Dynamic mechanism reduction is an on-the-fly mechanism reduction option where a reduced mechanism is generated for each computational cell [6]. A local reaction mechanism for any cell is significantly smaller than the original, larger mechanism.

It should be noted that many of the initial engine GA optimization studies that are described in this chapter were done before these acceleration strategies were available for engine CFD. Those studies are the main focus of this chapter as they laid the groundwork for the combined CFD and GA methodology. However, in the last decade or so, AMR (along with autonomous meshing), detailed chemistry acceleration techniques, and high-performance computing have proven to be key features for improving accuracy while keeping runtimes reasonable. As a result, the quality of optimization simulations done today is far superior to their predecessors, as illustrated by the last example described in Ref. [33]. This case used the CFD acceleration strategies described in this section.

2.2 Engine geometry generation

With the CFD code selected and validated, the next step is to define the input parameters of interest. Many inputs are relatively easy to adjust. For example, start of injection (SOI) timing is very straightforward, although care should be taken to also modify any parameters that are dependent on SOI, such as grid embedding or AMR settings.

On the other hand, including engine geometry in an optimization study can be much more difficult. The challenge becomes how to represent geometries using a set of parameters that can be assigned ranges. While full three-dimensional geometry optimization has certainly been carried out, it is much easier to illustrate the concept on a two-dimensional shape. This is the approach for diesel engines, for example, as their bowl shapes are commonly axisymmetric. We will now describe two parameterization techniques that have been applied to piston bowl geometry optimization.

2.2.1 Method of splines

In the "method of splines," the engine piston bowl profile is generated by means of a differentiable curve defined piecewise through polynomials. A cost-effective alternative consists of using a quadratic Bezier curve due to its relative simplicity but high flexibility. Fig. 4.2 shows an example where the

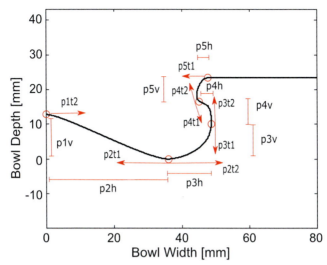

Figure 4.2 Parametrization of an engine piston bowl geometry. *(Credit: From Benajes J, Novella R, Pastor JM, Hernández-López A, Kokjohn SL. Computational optimization of a combustion system for a stoichiometric DME fueled compression ignition engine. Fuel 2018;223:20−31.)*

bowl profile contains five points known as nodes, and each node is related to two additional points (not necessarily over the bowl profile) which define the tension of the curve at that node.

Thus, in a two-dimensional problem, the number of parameters to account for increases by six for each node addition and then, additional relationships between these parameters are mandatory to decrease the number of optimization parameters down to affordable levels, keeping the desired degree of flexibility. This method was applied by Benajes et al. [38,39] to optimize the combustion system of a CI engine fueled with dimethyl ether (DME) and operating in lean or stoichiometric conditions.

2.2.2 Method of forces

In the so-called "method of forces," an engine piston bowl is parameterized as shown in Fig. 4.3. First, parameters are used to set the "center depth" and "bowl width." These parameters define the end points of the bowl profile. In between these points, a user-defined number of forces is specified (in Fig. 4.3, five forces are applied for a total of seven parameters). This can be thought of as applying forces to a rubber band. The forces are initially equidistant and act normal to the profile, given by a spline fit through the

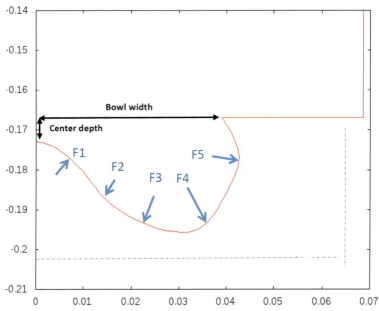

Figure 4.3 Illustration of the method of forces to parameterize an engine piston bowl.

points. The profile deflects, due to the forces, to an equilibrium position. This approach was used for the geometry optimization by Senecal et al. [17], which will be described later in this chapter.

2.3 Virtual injection model

An important input in the computational optimization of engine combustion systems is the injection rate profile, particularly in the case of CI engines operating with mixing-controlled combustion concepts due to the intrinsic relation between the injection and combustion processes.

The straightforward option consists of generating an experimental database containing a set of injection rate profiles as functions of the injection pressure, backpressure, injector energizing time duration, etc., and later generating the desired profiles by applying interpolation techniques.

A more refined option consists of adjusting a zero-dimensional model taking advantage of the information gathered in the experimental database. As shown in Fig. 4.4, this model is based on straight slopes for the injector needle opening and closing stages, a second order Bezier curve to soften the transition between the previous stages and the almost flat injection rate observed at maximum injector needle lift, together with a wave or logistic function to account for injection rate oscillations while the injector needle is fully lifted.

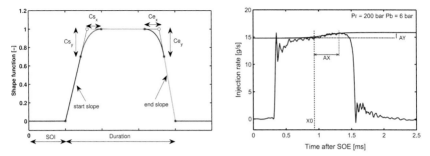

Figure 4.4 Straight slopes plus Bezier parameters (*left*). Logistic function parameters (*right*). *(Credit: From Payri R, Bracho G, Gimeno J, Bautista A. Rate of injection modelling for gasoline direct injectors. Energy Convers Manag 2018;166:424–432.)*

Further details about this promising zero-dimensional injection rate modeling approach are available in the literature for both gasoline (direct injection) [49] and diesel [50] injection rate simulations.

3. Optimization methods
3.1 Fundamentals of genetic algorithms

GAs follow the principles of evolution, where citizens in a population evolve over subsequent generations with successful characteristics passed on genetically to children. The mathematical procedure, first introduced by John Henry Holland in 1985, attempts to imitate natural evolution by creating a population of candidates, or generation of citizens, which are systematically subjected to a quality test. The best candidates are then selected to produce a new generation of citizens with improved traits. In addition, random variations of the best traits are incorporated to mimic aleatory genetic crossover and mutations also viewed in nature. This process, outlined in Fig. 4.5, is complete once a given citizen reaches sufficient quality after a certain number of generations [7].

Following the analogy of the genetic theory, the algorithm starts from a random initial population. Each individual represents a given chromosome which is a combination of multiple genes. Genes, as the minimum expression of genetic information, symbolize the input parameters to be optimized. Once each generation is computed, individuals are evaluated and ranked considering a fitness (or merit) function. Then, the subsequent generation is produced with the help of three genetic operators: selection, crossover, and mutation [8,9].

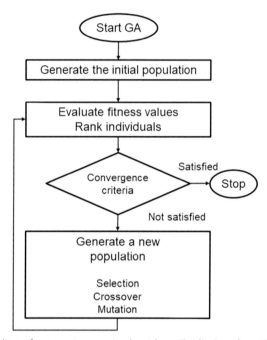

Figure 4.5 Outline of a generic genetic algorithm. *(Credit: Based on Hernández-López. Optimization and analysis by CFD of mixing-controlled combustion concepts in compression ignition engines [Ph. D. Thesis]: Univeristat Politècnica de València; 2018.)*

The selection operator mimics the concepts of natural selection and the "survival of the fittest" theory; the best individuals will become the parents of subsequent generations. In this sense, the best traits that emerge in each generation will be transferred to the future population.

The interaction between the genetic information of different individuals is imitated by the crossover operator. This operator tends to increase the average fitness of the new population since the attributes of fittest individuals are spread more frequently. Consequently, the similarity among individuals rises after several generations, reducing diversity and causing a stagnation of the population.

In order to compensate for this effect, the mutation operator increases the diversity of the population by introducing genetic alterations (by modifying some input parameters randomly) that can sometimes benefit future generations.

Differences between two different GAs reside in which mathematical approaches are used to define the genetic operators, despite sharing the same underlying framework. Thus, in practical and meaningful terms, the only difference lies in how the initial population evolves and finds the optimum.

3.2 Pioneering investigations

Looking at the literature, there is a significant delay between the mathematical formulation of the GA and its use for engine optimization. The lack of computational resources, that probably delayed its implementation to the ICE field for more than a decade, also conditioned the success of some particular approaches that aimed at modifying the existing algorithms to be applicable to computationally expensive problems. In this sense, the GA developed by Krishnakumar in 1989 [10] explored the advantages of minimizing the population size and the number of genetic parameters. His approach, coined the micro-genetic algorithm (μGA), demonstrated an excellent performance in solving nonstationary function problems with a reduced number of function evaluations. Due to this feature, its suitability for solving complex problems where the fitness function assessment consists of expensive and numerous parallel simulations is remarkable.

The application of GAs for engine optimization was pioneered by Senecal et al. [11]. In his seminal work [12], Senecal formulated a methodology for ICE optimization combining both multidimensional modeling (CFD) and the μGA approach proposed by Krishnakumar. He optimized a heavy-duty diesel engine using six engine input parameters related to injection (injection onset, split injection rate shape, and injection pressure) and gas exchange processes [amount of exhaust gas recirculation (EGR) and boost pressure]. The study investigated their effects on performance and pollutant emissions at a high-speed, medium-load operating condition using an objective, or merit, function to weight the contribution of NO_x, HC, and soot emission levels together with the fuel consumption. Besides, the optimized configuration predicted by computation was experimentally evaluated for validating the method and good correlation was found. In this sense, Senecal demonstrated the potential of this technique for the first time in ICE, proving its capacity for achieving reduced levels of pollutant emissions without performance penalties in a simultaneous optimization of a significant number of parameters.

Although this early effort was limited by computational resources, it already revealed two of the most relevant intrinsic virtues and shortcomings of GA-based optimization techniques. For instance, the existence of the optimum is strongly conditioned by the merit function and the characteristic weights applied to each objective parameter. In addition, he showed, through two executions of this optimization method, the issues of convergence and "local optima." Due to the random nature of the initial population, the optimum achievement is dependent on the number of realizations. Several authors tried justifying their optimum configurations by theoretically evaluating the method with algebraic functions that mimic the complex behavior of nature. However, the convergence of the method is not strictly verified using this method since, as other authors have shown [26–28], it strongly depends on the morphology of the evaluation function, which is unknown a priori. As in Senecal's work [12], other investigations [13] proposed a convergence criterion from the dispersion of each gene in a given population. For instance, the convergence of a particular gene is achieved if 95% of the individuals in the population have the same value for that gene. However, a certain user criterion is required, by the evolution of the fitness function, to determine when the optimum is reached.

This methodology was adapted and systematically applied in subsequent works [14–16]. Kim et al. [16] used a μGA for optimizing the boost pressure, the amount of EGR, and several injection parameters in a high-speed diesel engine. Similarly, Wickman et al. [14] optimized the combustion system design of two different diesel engines—a heavy-duty and a high-speed direct-injection (HSDI)—using this GA-based method with nine optimization parameters including three geometry-related variables. Being the first investigation to address aspects related to the combustion chamber geometry, their approach is rather simplistic. The complexity of modifying the bowl geometry while keeping the spray-oriented mesh configuration forced the authors to use a straightforward parametrization based on the bowl diameter, the bowl depth, and the central crown height of the piston. The results suggested that relatively large bowl diameters, long injections at high injection pressure through small nozzles, and moderate levels of swirl motion favor pollutant reduction while improving performance at medium-speed and high-load.

In view of the geometric limitations faced by Wickman et al. in Ref. [14], a novel parametrization approach for the piston bowl design was proposed by Senecal et al. [17]. They combined a new meshing strategy based on a hexahedral nonspray-oriented mesh with an external tool to generate the piston surface, called "the method of forces" which was described earlier in this chapter. However, the control over the bowl shape was not accurately resolved, which could lead to unrealistic geometries. Nevertheless, the variety of designs achieved with only six parameters was remarkable. Despite these limitations, this work closed the gap toward a fully flexible geometry generator for engine optimization.

It was Wickman in his seminal work [18], who proposed a new methodology for the bowl geometry parametrization that simplified the mesh generation by decoupling the geometry shape from the grid structure. The approach itself provided an enhanced flexibility since it allowed an independency of the CFD code mesh generation while assuring coherent and realistic geometries. The geometry generator module, *Kwickgrid*, used different parameters to create Bezier curves that deal with the desired piston bowl design, keeping the curvature details in a smooth way for real applications. This method was used in subsequent investigations [31,32] for evaluating the impact of different types of bowl geometries from more traditional reentrant and open-type shapes (axisymmetric) to new and more exotic ones like stepped-bowl geometries [19]. Despite using different geometrical relationships and mathematical criteria, this approach was also used in more recent investigations to generate extremely flexible bowl geometry shapes based on 15 [34] and 5 [33] input parameters.

The effect of different operating conditions in the optimization procedure was firstly evaluated by Senecal et al. [17] and Shrivastava et al. [15]. The former used the KIVA-GA methodology to optimize engine emissions and performance simultaneously for two operating conditions. The latter included the optimization of eight input parameters at four different operating conditions ranging from low-to-medium speed and low-to-high load. They showed, through the different engine requirements (i.e., flow field aspects are dominant at high-speed low-load conditions), that optimization criteria must be adapted to the operating condition, hindering the existence of a global optimum for the whole engine map.

Subramaniam and Reitz [13] applied this methodology for optimizing a two-stroke direct-injection engine operated under the spark-ignition concept using gasoline fuel. This investigation focused on reducing pollutant emissions and wall heat transfer (HT) while improving the fuel economy through the operating parameters at part load conditions. Although this work does not entail a significant advance in terms of the optimization methodology itself, it included two interesting aspects on the modeling side. They applied the linearized instability sheet atomization model [20] for simulating the breakup phase of the fuel injection process. In addition, they implemented the discrete particle ignition kernel model of

Fan et al. [21] for mimicking the energy deposition of the spark arc. The results also demonstrated the capability of evolutionary-driven multidimensional simulations for engine performance improvement in such specific application. In this sense, this work was the earliest in the optimization of SI-based combustion systems, although it did not include any aspect related to the geometric design of the combustion chamber.

3.3 Multiobjective framework

Subsequent studies introduced the concept of multiobjective GA-based optimization (MOGA) [22–24]. All of the investigations described so far are simplified cases of multiobjective optimization problems that were artificially converted into single objective problems via a multicomponent fitness function. But in fact, the heuristic method and the algorithm structure used for determining the optimum in these two classes of evolutionary methods are considerably different. For single objective formulations, there only exists a unique, obvious, and clear optimum solution at each stage of the evolutionary process. On the contrary, for multiobjective optimization problems, the different objectives can contradict each other resulting in a pool of optimal solutions, denoted as a Pareto front, which should be tracked by the algorithm during the evolution. These fundamental aspects do not necessarily condition the achievement of the best solution for a given problem, but they have an important impact on the information, in the widest sense, which can be extracted from the optimization process. The most interesting aspect is probably the one depicted in Fig. 4.6 by Lee and Park in Ref. [44]. Data obtained from a multi-objective optimization not only help to identify an optimum solution that minimizes (or maximizes) the global objective but also allow for the inspection of how this solution changes as one (or more) of the objectives prevails over the other ones. For instance, it can be seen in the same figure how the system configuration changes from a slightly reentrant bowl geometry (E) when all objectives are minimized (NO_x, soot and GISFC) to a high reentrant bowl geometry (B) when GISFC prevails over both soot and NO_x emissions. This is an important aspect of these kinds of optimization procedures where extracting the maximum amount of information is imperative due to their large computational cost.

After introducing this variant of GA, several options emerged as new potential strategies for engine combustion systems optimization. Shi and Reitz [25] evaluated the performance of three different multiobjective GAs

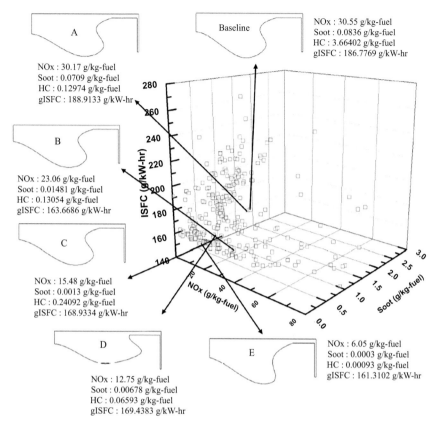

Figure 4.6 Visualization of the Pareto surface solution for the optimization of a piston bowl geometry. Differences in the configuration are highlighted as one of the objectives prevails over the other two. *(Credit: From Lee S, Park S. Optimization of the piston bowl geometry and the operating conditions of a gasoline-diesel dual-fuel engine based on a compression ignition engine. Energy 2017;121:433–448.)*

for conducting the optimization of a heavy-duty diesel engine under the same boundary conditions. The study included the μGA, initially developed by Krishnakumar [10] for SOGA problems and later adapted to MOGA by Coello and Pulido [26], the nondominated sorting genetic algorithm II (NSGA-II) presented by Deb et al. [27], and the adaptive range multiobjective genetic algorithm (ARMOGA) by Sasaki and Obayashi [28] as the most widely used methods in engineering applications. In addition, they considered two scenarios with different population sizes (4 and 32) for the two latter algorithms. The bases of the optimization are characterized by seven input parameters (five related to the chamber geometry, the injector included spray angle, and the swirl flow intensity) and three simultaneous objectives: fuel economy, NO_x, and soot emissions. The algorithms were ranked by different parameters. The number of Pareto solutions accounts for the number of optimal designs available for the optimum selection. The mean distance to the Pareto front (MDPF) represents the proximity of the Pareto solutions predicted by the MOGA to the true Pareto front. Similarly, the mean distance between extreme Pareto solutions (MDEPS) indicates how far the predicted Pareto is from the extreme values of each objective. The mean deviation of the distance between neighbor Pareto solutions (MDDNPS) grades the distribution of the optimal space, estimating the quality of the database obtained after the optimization for further studies based on regression methods.

Results showed that MOGAs with larger populations give more Pareto solutions than those with smaller ones. In particular, NSGA II with a large population number (32 individuals) generates the most Pareto solutions even with a small number of evaluations. A large initial population increases the probability of finding those traits needed for optimal designs. However, only increasing the initial population does not guarantee more Pareto solutions, as seen for μGA, since the merits of the initial individuals should be preserved in the subsequent generations. Moreover, the same study highlights how the distance to the true Pareto front is reduced as the number of generations increase. Results of μGA and ARMOGA with small populations (4) reach smaller values of MDPF compared to NSGA II and ARMOGA with large population sizes (32). Poor performance of NSGA II with a small population size suggests that selection of the population size is a critical aspect for this algorithm.

The ability of MOGAs to extend the limits of their solution space is represented by the MDEPS parameter. The algorithms with a large population tend to outperform the ones with small populations. In general, the enhanced performance of NSGA II (32) was evident as it gives high values of MDEPS along the whole evolutionary process.

Regarding how well distributed the optimal solutions along the Pareto front are, MDDNPS calculated in Ref. [25] evince again the good performance of NSGA II, especially if a large population is considered, compared to other MOGAs. The low values of MDDNPS reveal a suitable spreading of the optimal solutions, resulting in a good database for further analyses based on regression methods.

In view of these results, the authors concluded that the NSGA II with an adequate population size is the best option for highly nonlinear

multiobjective problems, such as combustion system optimization in ICEs. Following the original principle of GAs, survival of the fittest, the NSGA II was motivated by the idea of promoting a better convergence in multi-objective evolutionary optimization with the help of the concept of elitism. The algorithm combines a strategy based on the elite preservation with an explicit diversity propagation mechanism. The evolutionary process of NSGA II evolves under conventional selection, crossover, and mutation directives from a first randomly distributed population. During this process, the parent population transmits its traits to the next generation introducing elitism through two main rules applied to the selection operator:
1. Individuals better ranked have an increased probability of transmitting its traits.
2. For the individuals with the same rank, a higher priority is assigned for the less crowding distance one.

Since one of the best features of NSGA II is the spreading of the Pareto solutions, the suitability of this data for improving the understanding of each operating parameter over the objectives of the optimization is evident. For this reason, several authors evaluated the performance of regression methods to post-process results of these optimal solutions and to shed some light on the connection between the combustion system design and both fuel economy and pollutant formation.

Fig. 4.7 shows an example of this type of study, in which Kriging and NN regression methods were utilized in Ref. [48] to visualize the interferences of two design parameters (start and duration of injection) with the optimization objective (merit value). As can be seen, both methods provide similar qualitative results and the relationships between the optimization parameters and the objectives are easily identified. For instance, it

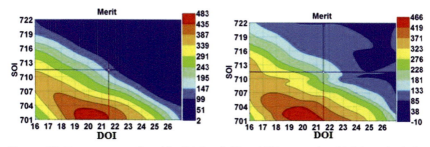

Figure 4.7 Responses predicted by Kriging (*left*) and NN regression (*right*) methods for a merit function value. *(Credit: Data from [48].)*

can be observed that the impact of DOI is higher than SOI on the merit value, suggesting that this parameter can be used as an additional degree of freedom for optimizing secondary objectives.

The use of these latter techniques promoted a paradigm shift that favored the increase of CFD evaluations in an optimization procedure. While, in the first optimizations, the core idea was to find an optimal solution with the minimum number of function evaluations, the appearance of these studies and their inherent advantages with regard to the understanding of the key aspects diffused this goal. Thus, the number of CFD simulations in optimization procedures increased systematically, becoming a standard from 2009 to 10 onwards. This transition, evinced in Table 4.1, was also favored by the increase in computational performance and resources, which made it possible to increase the number of parameters to be optimized significantly. Shi and Reitz [30] optimized 10 input parameters at two different engine load conditions, with more than 1000 simulations performed. In subsequent investigations [31,32], Ge, Shi, and Reitz et al. increased up to 15 optimization parameters, 32 individuals per population, and more than 2200 CFD evaluations in a full optimization procedure.

Once the methodology was validated and its basis well understood, several authors started to use it in other applications within the field of ICE. As done by Subramaniam and Reitz in a two-stroke gasoline engine [13], some works focused on the optimization of the combustion system of conventional SI engines. For example, Sun et al. [35] optimized the flow performance, that is the tumble motion, in an SI engine. They proposed a parametrization of the intake port by 18 predefined parameters to create different structural configurations that help to increase the tumble motion, promoting flame propagation and thus, increasing the burning rate. Results evinced a 6.12% gain in terms of tumble motion whereas the flow coefficient was kept constant.

In the field of CI diesel engines, in which the CFD-GA optimization was originally developed, research focused on evaluating different strategies to reduce pollutant formation. With the emergence of new combustion concepts based on low temperature combustion, CFD-GA optimizations were postulated as the perfect tool for optimizing the combustion system and for understanding its requirements. In this context, Shi and Reitz [41] optimized the combustion system of a heavy-duty CI engine operating with gasoline fuel (GCI). Researchers at the University of Wisconsin–Madison in collaboration with Oak Ridge National Laboratory optimized the RCCI in a heavy-duty [42] and a light-duty engine [43].

Table 4.1 Summary of some of the most relevant investigations related to computational fluid dynamics-genetic algorithms.

Authors	GA	GA mode	CFD code	Engine (fuel)	Opt. Params.	Pop. Size	Evals.	Year
Senecal et al. [11]	μGA	SOGA	KIVA-3V	Heavy-duty CI (diesel)	6	5	500	2000
Wickman et al. [14]	μGA	SOGA	KIVA-3V	Heavy-duty CI (diesel)	9	5	400	2001
				HSDI small-bore (diesel)			400	
Shrivastava et al. [15]	μGA	SOGA	KIVA-3V	Medium-duty (diesel)	8	5	380	2002
Senecal et al. [17]	μGA	SOGA	KIVA-3V	DI small-bore (diesel)	12	9	1072	2002
Subramaniam et al. [13]	μGA	SOGA	KIVA-3V	Two-stroke DI (gasoline)	6	5	425	2003
Kim et al. [16]	μGA	SOGA	KIVA-3V	Heavy-duty CI (diesel)	5	5	750	2005
Genzale et al. [24]	μGA	MOGA	KIVA-3V	Heavy-duty CI (diesel)	9	4	800	2007
Shi and reitz [25]	μGA NSGA II ARMOGA	MOGA	KIVA-3V	Heavy-duty CI (diesel)	7	4 4 | 32 4 | 32	1280	2008

Continued

Table 4.1 Summary of some of the most relevant investigations related to computational fluid dynamics-genetic algorithms.—cont'd

Authors	GA	GA mode	CFD code	Engine (fuel)	Opt. Params.	Pop. Size	Evals.	Year
Shi and reitz [30]	NSGA II	MOGA	KIVA-3V rel. 2	Heavy-duty CI (diesel)	10	24	1272	2008
Shi and reitz [41]	NSGA II	MOGA	KIVA-3V rel. 2	Heavy-duty CI (gasoline)	8	32	800	2010
Dolak et al. [19]	NSGA II	MOGA	KIVA-3V rel. 2	Light-duty (diesel)	15	32	1376	2010
Lee and Park [44]	μGA	SOGA	KIVA-3V	Light-duty (dDual-fuel)	16	6	1620	2017
Benajes et al. [34]	DKGA	SOGA	KIVA-3V	Heavy-duty CI (DME)	22	529	15,879	2018
Broatch et al. [33]	DKGA	SOGA	CONVERGE v2.2.17	Light-duty CI (diesel)	8	25	725	2018
Zubel et al. [47]	μGA (Congo)	SOGA	CONVERGE v2.3.6	Light-duty CI (DME)	7 8	7 8	700 800	2019

μGA, *micro-genetic algorithm*; *CI*, compression ignition; *DME*, dimethyl ether; *HSDI*, high-speed direct-injection.

3.4 Convergence acceleration

Although MOGAs are interesting for research and scientific understanding, SOGAs have certain advantages from an industrial and resources point of view. For this reason, other authors followed the latter path to find optimal solutions in ICE applications, and/or to include other relevant aspects within the optimization process. The so-called DKGA [36] was developed in order to overcome some of the issues found in the application of previously described algorithms to ICE optimization. This algorithm has two main differences: the chromosomes are represented in decimal format, and the initial mutations are large but decrease while the optimization progresses to force convergence. The algorithm workflow can be summarized in the following steps:

1. Select the best designs to be the parents for the next generation. For the first generation, parents are the initial conditions provided by the user.
2. Crossbreed the parents using the Punnett square technique (discussed below) to create a new generation.
3. Mutate each chromosome of each child in the generation.
4. Test the population against a fitness function.
5. Penalize children that surpass the constraints.
6. Sort the population from highest to lowest.
7. Repeat from step one until the maximum number of generations are complete.

In the DKGA, the mating selection is performed by using a Punnett square where the top n designs from the previous generations become the parents of the new generation and each parent has a child with every other parent twice and themselves once producing a new generation of n^2 children. After the new generation is created, each variable of each child is then mutated. A normally distributed random number with its mean set to the current value and standard deviation set by a decaying time constant is generated and adds mutations to the system. As the GA progresses, the time constant that dictates the mutation rate exponentially decays. The rate at which it decays is defined as follows:

$$\tau_{GA,i} = \tau_{GA,0} * \exp\left(-\sigma_{GA}\frac{i}{n_{end}}\right)$$

where $\tau_{GA,i}$ is the time constant at the ith generation, $\tau_{GA,0}$ is the user specified initial time constant, σ_{GA} is the standard deviation, and n_{end} is the user specified total number of generations. The decaying time constant enables coverage of the full design space in early generations and forces GA convergence in later generations.

It was shown by Klos [36] and Hernández-López [37] that DKGA outperforms both μGA and NSGA-II on a benchmark study using a fitness function. This function, described by the following equation where s is the number of variables, presents several local optima that make it difficult to find the global optimal solution.

$$FitnessFunc = \prod_{i=1}^{S} |15 * X_i * (1 - X_i) * sin(n * \pi * X_i)|$$

Fig. 4.8 depicts an example where DKGA finds the optimum on most of the repetitions after a given number of function evaluations, while both μGA and NSGA-II converge slowly or even are not able to obtain this fitness function's optimum.

The DKGA was successfully applied by Benajes et al. [34] for the optimization of a heavy-duty engine operating with DME, an alternative oxygenated nonsooting fuel. The optimization aimed at maximizing the net indicated efficiency (NIE) while keeping pollutants, in-cylinder peak pressure (PP) and rate below specified limits. The inputs comprised 22 parameters, including piston bowl geometry, nozzle design, as well as injection and air management settings. A code capable of generating an arbitrarily shaped axisymmetric piston bowl geometry, and automatically producing a suitable mesh for the KIVA-3V code [40], was developed. The bowl shape was parametrized by five control points, where each control

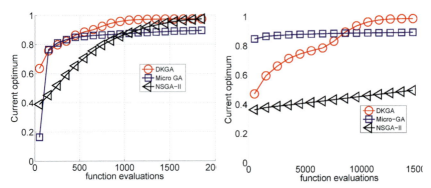

Figure 4.8 Optimization results for the fitness function considering three different genetic algorithms. *(Credit: From [39].)*

point was connected by a Bezier spline. The study used a turbocharger model to calculate the pumping work needed to achieve the required intake boost, and to obtain the net indicated work.

This methodology was further applied to comply for US 2010 emission standards [38], and to US 2030 regulations by operating the combustion system under stoichiometric conditions, coupled to a three-way catalyst [39]. These works showed the potential of DKGA for the optimization of nonconventional combustion systems, where a high number of design parameters and ranges are present. Both optimizations were run for 40 generations with a population size of 529 cases per generation, resulting in a total of 21,160 function evaluations.

The evolution of the optimization process published in Ref. [39] and how all the parameters converged to the optimum configuration, are shown in Fig. 4.9. The first parameters to converge were the EGR and the

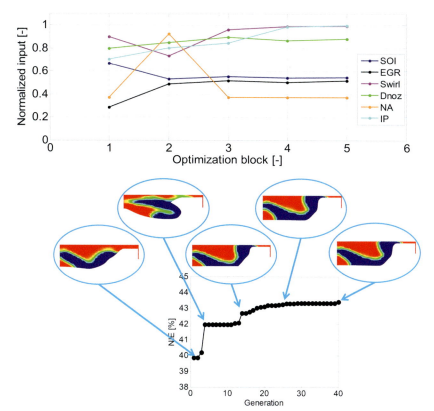

Figure 4.9 Evolution of the optimization parameters and optimal solutions. *(Credit: From [38].)*

SOI. These two parameters were critical to control PP and NO_x and then, the parameters that have the most effect over the constraint outputs converged earlier in the optimization. The piston bowl geometry and nozzle include angle (NA) were the next parameters to converge. In this case, the optimization process switched between reentrant and nonreentrant shapes, made possible by the capability of the algorithm to follow two or more optimization paths until one path overcomes the rest. After 13 generations, the nonreentrant shapes were able to offer higher NIE results and the algorithm rapidly converged to that type of geometry. Finally, once the geometry was set and the NO_x emissions and pressure were controlled, the algorithm improved the NIE with the injection settings until the end of the process.

Fig. 4.10 compares the optimum piston shape with the baseline geometry. The process shifted the reference piston geometry toward a nonreentrant shape. The optimum geometry is flat (i.e., the GA removed the pip) and shallower than the stock diesel bowl. The piston width is similar to the baseline engine. Additionally, the NA was selected to be slightly narrower than the baseline geometry.

The optimization results were analyzed using the nonparametric regression model based on the COSSO method [29], not only to provide the optimal configuration but also to identify the pathways to a high-efficiency DME fueled engine. Fig. 4.11 shows the impact of EGR, swirl

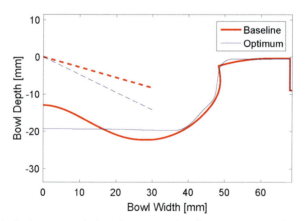

Figure 4.10 Optimum and baseline case bowl geometry and nozzle angle configuration. *(Credit: From [39].)*

Engine optimization using computational fluid dynamics and genetic algorithms 95

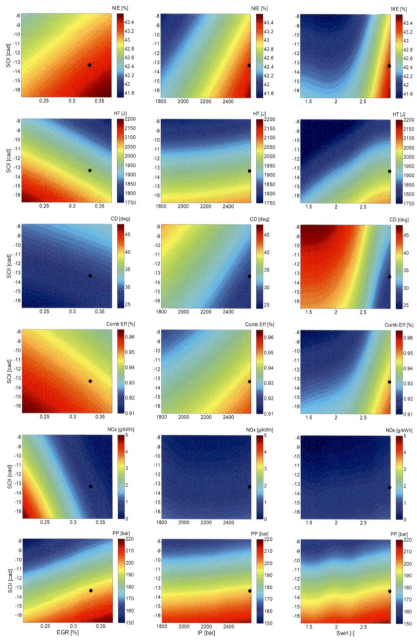

Figure 4.11 Response surfaces obtained for visualization of the coupled effect of exhaust gas recirculation (EGR), IP, Swirl with start of injection (SOI) over net indicated efficiency (NIE), heat transfer, combustion duration (CD), combustion efficiency, NO$_x$, and peak pressure (PP). The optimum value of a given input is highlighted by the black dot. *(Credit: From [39].)*

ratio, IP, and SOI timing on performance [NIE, wall HT, combustion duration (CD), combustion efficiency (CE), and PP] and NO_x emissions. The results show that when a stoichiometric mixture is maintained, the trade-off between NO_x and NIE is removed and both parameters improve with increasing EGR. Higher EGR levels reduce the oxygen mole fraction, reducing the adiabatic flame temperature and decreasing NO_x emissions and peak temperature. Additionally, at a fixed equivalence ratio, higher EGR levels increase the in-cylinder trapped mass, further reducing the in-cylinder temperature and HT losses. Conversely, since the oxygen concentration is reduced, lack of free oxygen results in slower combustion (increased burn duration) and lower CE. Nevertheless, the reduced HT losses outweigh the longer burn duration and lower CE, resulting in a net increase in NIE with increased EGR. The effect of IP is similar to that of swirl ratio, that is, both increase HT, but also shorten CD resulting in an NIE improvement. It is interesting to note that IP and swirl ratio are complimentary. This shows the importance of the mixing process in stoichiometric DI engines.

An interesting example of the DKGA approach was performed by Broatch et al. [33], where the optimization was focused on both efficiency and noise control for an HSDI engine. The CONVERGE code [3] was used to perform the CFD calculations, which enables better resolved and even grid-convergent simulations with runtimes compatible with these optimizations. They also considered such relevant and complex aspects related to high frequency noise emission. As shown in Fig. 4.12, the optimized combustion system was able to reduce noise emissions, owing to the lowering of resonance energy, since it modified the frequency content to feature higher frequencies, less perceptible by human hearing.

This computational study demonstrated a comprehensive framework to incorporate combustion noise into a numerical optimization strategy for engine design. A modal decomposition analysis based on proper orthogonal decomposition was included in the investigation to establish relationships between the optimizing parameters and the spectral response of the resulting pressure field. The visualization of the most relevant oscillation patterns included in Fig. 4.13 revealed the connection between the high frequency components and their associated oscillation patterns. In addition, oscillations focused within the bowl gather higher frequencies with lesser sound pressure contribution that helped to decrease the overall noise emission.

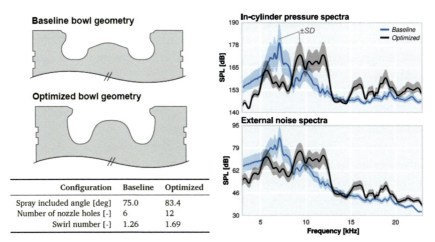

Figure 4.12 Comparison of the baseline and optimized configurations (Left). Comparison of the in-cylinder pressure spectra trends (Right). The pressure spectrum averaged over all cells in the domain is plotted with its standard deviation (±SD) for both configurations (baseline and optimized). *(Credit: Modified from [33].)*

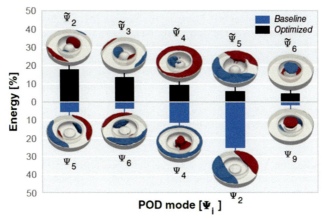

Figure 4.13 Energy share and spatial distribution of the five most relevant optimized design oscillation patterns (top), together with their most closely resembling baseline counterparts (bottom). *(Credit: From [33].)*

4. Summary and concluding remarks

The optimization of ICEs has experienced a breakthrough with the implementation of numerical methods that couple CFD with evolutionary methods (CFD-GA) in the engine design and calibration processes.

The different optimization approaches demonstrated, through distinct methods, that this combination helps to identify potential configurations and design paths in the search for higher efficiency and low pollutant emissions.

The development of new calculation techniques, such as adaptive meshing and detailed chemistry acceleration, for speeding up the simulations and for keeping the accuracy of the results allowed the implementation of CFD in large-scale and massive engine simulations. Moreover, advances in the investigation of GAs increased the flexibility of the optimization procedure itself by enhancing the exploration of the optimization-space. For instance, while some of these methods specifically focus on achieving the optimum configuration as quickly as possible (i.e., μGA and DKGA), others try to generate the maximum data to improve the understanding of the optimization space and to propose some actions to improve the engine design concept (i.e., NSGA II).

References

[1] Baden SB, Chrisochoides NP, Gannon DB, Norman ML, editors. Structured adaptive mesh refinement (SAMR) grid methods. New York: Springer-Verlag; 2000.
[2] Senecal P, Richards K, Pomraning E, Yang T, et al. A new parallel cut-cell Cartesian CFD code for rapid grid generation applied to in-cylinder diesel engine simulations. SAE Technical Paper 2007-01-0159. 2007.
[3] Richards, K. J., Senecal, P. K., Pomraning, E., "Converge manual, Convergent Science, Inc., Madison, WI (2008−2021).
[4] Raju M, Wang M, Dai M, Quan S, Senecal PK, Sibendu Som, McNenly MJ, Flowers D. Towards accommodating comprehensive chemical reaction mechanisms in practical internal combustion engine simulations. In: 8th U. S. National combustion meeting, may 19-22; 2013.
[5] Babajimopoulos A, Assanis D,N, Flowers DL, Aceves SM, Hessel RP. A fully coupled computational fluid dynamics and multi-zone model with detailed chemical kinetics for the simulation of premixed charge compression ignition engines. Int J Engine Res 2005;6(5):497−512.
[6] Raju M, Wang M, Senecal PK. Dynamic chemical mechanism reduction for internal combustion engine simulations. SAE 2013-01-1110. 2013.
[7] Eiben AE, Smith JE. Genetic algorithms. In: Introduction to evolutionary computing. Berlin, Heidelberg: Springer; 2003. p. 37−69.
[8] Kachitvichyanukul V. Comparison of three evolutionary algorithms: GA, PSO, and DE. Industrial Engineering and Management Systems 2012;11(3):215−23.
[9] Hassan R, Cohanim B, De Weck O, Venter G. A comparison of particle swarm optimization and the genetic algorithm. In: 46th AIAA/ASME/ASCE/AHS/ASC structures, structural dynamics and materials conference; April 2005. p. 1897.
[10] Krishnakumar K. Micro-genetic algorithms for stationary and non-stationary function optimization. Intelligent control and adaptive systems, vol. 1196. International Society for Optics and Photonics; February 1990. p. 289−96.

[11] Senecal PK, Montgomery DT, Reitz RD. A methodology for engine design using multi-dimensional modelling and genetic algorithms with validation through experiments. Int J Engine Res 2000;1(3):229−48.
[12] Senecal PK. Development of a methodology for internal combustion engine design using multi-dimensional modeling with validation through experiments. 2000.
[13] Subramaniam MN, Reitz RD, Ruman M. Reduction of emissions and fuel consumption in a 2-stroke direct injection engine with multidimensional modeling and an evolutionary search technique. SAE Trans 2003;2003-01-0544:718−36.
[14] Wickman DD, Senecal PK, Reitz RD. Diesel engine combustion chamber geometry optimization using genetic algorithms and multi-dimensional spray and combustion modeling. SAE Trans 2001;2001-01-0547:487−507.
[15] Shrivastava R, Hessel R, Reitz RD. CFD optimization of DI diesel engine performance and emissions using variable intake valve actuation with boost pressure, EGR and multiple injections. SAE Trans 2002;2002-01-0959:1612−29.
[16] Kim M, Liechty MP, Reitz RD. Application of micro-genetic algorithms for the optimization of injection strategies in a heavy-duty diesel engine (No. 2005-01-0219). SAE Technical Paper. 2005.
[17] Senecal PK, Pomraning E, Richards KJ. Multi-mode genetic algorithm optimization of combustion chamber geometry for low emissions (No. 2002-01-0958). SAE Technical Paper. 2002.
[18] Wickman DD. HSDI diesel engine combustion chamber geometry optimization. 2004.
[19] Dolak JG, Shi Y, Reitz RD. A computational investigation of stepped-bowl piston geometry for a light duty engine operating at low load (No. 2010-01-1263). SAE Technical Paper. 2010.
[20] Senecal PK, Schmidt DP, Nouar I, Rutland CJ, Reitz RD, Corradini ML. Modeling high-speed viscous liquid sheet atomization. Int J Multiphas Flow 1999;25(6−7):1073−97.
[21] Fan L, Li G, Han Z, Reitz RD. Modeling fuel preparation and stratified combustion in a gasoline direct injection engine. SAE Trans 1999;1999-01-0175:105−19.
[22] De Risi A, Donateo T, Laforgia D. Optimization of the combustion chamber of direct injection diesel engines. SAE Trans 2003;2003-01-1064:1437−45.
[23] De Risi A, Donateo T, Laforgia D. An innovative methodology to improve the design and the performance of direct injection diesel engines. Int J Engine Res 2004;5(5):425−41.
[24] Genzale CL, Reitz RD, Wickman DD. 2007-01-0119 A computational investigation into the effects of spray targeting, bowl geometry and swirl ratio for low-temperature combustion in a heavy-duty diesel engine. SAE SP 2007;2082:1.
[25] Shi Y, Reitz RD. Assessment of optimization methodologies to study the effects of bowl geometry, spray targeting and swirl ratio for a heavy-duty diesel engine operated at high-load. SAE International Journal of Engines 2008;1(1):537−57.
[26] Coello CACC, Pulido GT. A micro-genetic algorithm for multiobjective optimization. In: International conference on evolutionary multi-criterion optimization. Berlin, Heidelberg: Springer; March 2001. p. 126−40.
[27] Deb K, Agrawal S, Pratap A, Meyarivan T. A fast elitist non-dominated sorting genetic algorithm for multi-objective optimization: NSGA-II. In: International conference on parallel problem solving from nature. Berlin, Heidelberg: Springer; September 2000. p. 849−58.

[28] Sasaki D, Obayashi S. Adaptive range multi-objective genetic algorithms and self-organizing map for multi-objective optimization problem. VKI lecture series: optimization methods & tools for multicriteria/multidisciplinary design, applications to aeronautics and turbomachinery. Belgium: Rhode-Saint-Genese; 2004.
[29] Liu Y, Lu F, Reitz RD. The use of non-parametric regression to investigate the sensitivities of high-speed direct injection diesel emissions and fuel consumption to engine parameters. Int J Engine Res 2006;7(2):167−80.
[30] Shi Y, Reitz RD. Optimization study of the effects of bowl geometry, spray targeting, and swirl ratio for a heavy-duty diesel engine operated at low and high load. Int J Engine Res 2008;9(4):325−46.
[31] Ge HW, Shi Y, Reitz RD, Wickman DD, Willems W. Optimization of a HSDI diesel engine for passenger cars using a multi-objective genetic algorithm and multi-dimensional modeling. SAE International Journal of Engines 2009;2(1):691−713.
[32] Ge HW, Shi Y, Reitz RD, Wickman D, Willems W. Engine development using multi-dimensional CFD and computer optimization (No. 2010-01-0360). SAE Technical Paper. 2010.
[33] Broatch A, Novella R, Gomez-Soriano J, Pal P, Som S. Numerical methodology for optimization of compression-ignited engines considering combustion noise control. SAE International Journal of Engines 2018;11(6):625−42.
[34] Benajes J, Novella R, Pastor JM, Hernández-López A, Kokjohn SL. Computational optimization of the combustion system of a heavy duty direct injection diesel engine operating with dimethyl-ether. Fuel 2018;218:127−39.
[35] Sun Y, Wang T, Lu Z, Cui L, Jia M. The optimization of intake port using genetic algorithm and artificial neural network for gasoline engines (No. 2015-01-1353). SAE Technical Paper. 2015.
[36] Klos DT. Investigations of low temperature combustion (LTC) engine design and combustion instability. M.S. Thesis. Madison: University of Wisconsin; 2015.
[37] Hernández-López. Optimization and analysis by CFD of mixing-controlled combustion concepts in compression ignition engines. Ph. D. Thesis. Univeristat Politècnica de València; 2018.
[38] Benajes J, Novella R, Hernández-López A, Kokjohn SL. Numerical optimization of the combustion system of a HD compression ignition engine fueled with DME considering current and future emission standards (No. 2018-01-0247). SAE Technical Paper. 2018.
[39] Benajes J, Novella R, Pastor JM, Hernández-López A, Kokjohn SL. Computational optimization of a combustion system for a stoichiometric DME fueled compression ignition engine. Fuel 2018;223:20−31.
[40] Amsden A. KIVA-3V, release 2, Improvments to KIVA-3V. LA-UR-99-915. 1999.
[41] Shi Y, Reitz RD. Optimization of a heavy-duty compression−ignition engine fueled with diesel and gasoline-like fuels. Fuel 2010;89(11):3416−30.
[42] Nieman DE, Dempsey AB, Reitz RD. Heavy-duty RCCI operation using natural gas and diesel. SAE International Journal of Engines 2012;5(2):270−85.
[43] Hanson R, Curran S, Wagner R, Kokjohn S, Splitter D, Reitz R. Piston bowl optimization for RCCI combustion in a light-duty multi-cylinder engine. SAE International Journal of Engines 2012;5(2):286−99.
[44] Lee S, Park S. Optimization of the piston bowl geometry and the operating conditions of a gasoline-diesel dual-fuel engine based on a compression ignition engine. Energy 2017;121:433−48.
[45] Crowley J, Hoering A, editors. Handbook of statistics in clinical oncology. CRC Press; 2012.

[46] Antony J. Design of experiments for engineers and scientists. Elsevier; 2014.
[47] Zubel M, Ottenwälder T, Heuser B, Pischinger S. Combustion system optimization for dimethyl ether using a genetic algorithm. Int J Engine Res 2021;22(1):22—38.
[48] Jafari M, Parhizkar MJ, Amani E, Naderan H. Inclusion of entropy generation minimization in multi-objective CFD optimization of diesel engines. Energy 2016;114:526—41.
[49] Payri R, Bracho G, Gimeno J, Bautista A. Rate of injection modelling for gasoline direct injectors. Energy Convers Manag 2018;166:424—32.
[50] Payri R, Gimeno J, Novella R, Bracho G. On the rate of injection modeling applied to direct injection compression ignition engines. Int J Engine Res 2016;17(10):1015—30.

CHAPTER 5

Computational fluid dynamics—guided engine combustion system design optimization using design of experiments

Yuanjiang Pei[1], Anqi Zhang[1], Pinaki Pal[2], Le Zhao[2], Yu Zhang[1] and Sibendu Som[2]

[1]Aramco Americas: Aramco Research Center—Detroit, Novi, MI, United States; [2]Energy Systems Division, Argonne National Laboratory, Lemont, IL, United States

1. Introduction

With ever-rising demand for better fuel economy and lower criteria emissions, the automotive industry has been pushing the exploration of novel designs and combustion strategies to improve the efficiency and reduce the emissions of internal combustion engines (ICEs). However, engine performance is highly sensitive to a large number of parameters associated with the fuel type, combustion system design, fuel injection system, air system, operating conditions, and their complex interactions. The number of parametric tests dramatically increases as the number of input parameters and their variation levels increase, which renders fuel-engine co-optimization within such a huge design space as a challenging task. Therefore, it is very important to reduce the number of required tests (experimental or computational) without loss of accuracy in determining the optimum engine design or calibration.

Studying the design parameters one at a time or by trial and error until a first feasible design is found is a commonly practiced approach. However, this leads either to a lengthy and costly time span or to a premature termination of the design process due to cost or schedule constraints. As a result, the design may be far from optimal. In contrast, design of experiments (DoE) is a structured and organized method in which planned variations are made to the input variables of a system, and the effects of these variations on outputs are analyzed. In short, it allows for discovering cause-effect relationships. The method of analysis is to look for differences between response (output) readings for different groups of the input changes. These differences are then attributed to the input variables acting alone (i.e., single effect) or in combination with another input variable (i.e., interaction). Two types of DoE methods are generally used: full factorial and fractional factorial methods [1]. The full factorial method is the most comprehensive method, where all possible interactions between the input variables are considered, but it is limited to only a small number of input variables, since the number of experiments increases exponentially with the number of variables. The fractional factorial method, on the other hand, is a more practical method that offers the potential to reduce the number of experiments while still retaining sufficient fidelity to capture the response from important design variables and the first-order interactions.

First, it is essential to define proper objective functions to achieve clear and quantitative indications during engine calibration/optimization. The search for the optimum configuration must be performed on the response surface defined on the hyperspace of the input parameters. For complex systems like ICEs, high-fidelity statistical response models need to be created by using measured or computed outputs. Second, it is worth noting that this is an interactive and iterative process. The construction of the response model is based on the results from experimental/numerical tests. These tests need to be carefully planned and carried out to attain good confidence that the optimal design is contained within the defined boundary conditions. The search domain is also subjective to adjustment during the optimization process by observing the outputs. Finally, the response surface model (RSM) and the subsequent optimization method need to be carefully evaluated and selected so that the most appropriate approach is used.

Over the years, the DoE optimization method has been widely used for engine calibration and design optimization. Besson et al. [2] optimized an homogeneous charge compression ignition (HCCI) combustion chamber operating at full load using computational fluid dynamics (CFD)-DoE. Han et al. [3] used DoE to optimize the intake and spray injection patterns of a new light stratified-charge direct injection (DI) spark-ignition (SI) combustion system. Mallamo et al. [4] coupled DoE with experiments to optimize multiple injection strategies for low emissions, combustion noise and brake specific fuel consumption in a small-displacement Common Rail off-road diesel engine. Lippert et al. [5] optimized piston and injector design of a small displacement SI DI engine under stratified operation using CFD-DoE comprising three piston designs and seven injector configurations. Hajireza et al. [6] used a DoE size of 16 designs to optimize the piston bowl, spray cone angle, and intake port swirl level of a diesel engine using CFD. Gothekar et al. [7] carried out an optimization of a naturally aspirated DI diesel engine using DoE for optimum performance and minimum emissions. Davis et al. [8] developed a combustion system for the 3.6L DOHC 4V V6 DI engine using the CFD-DoE optimization approach. DoE was employed by Reiche et al. [9] to experimentally optimize the cold start for the EcoBoost engine. Catania et al. [10] used experimental DoE to optimize a premixed charge compression ignition combustion system. Styron et al. [11] used the CFD-DoE approach to optimize the Ford 2011 6.7L Power Stroke diesel engine with a DoE size of 16 designs comprising three pistons. Rajamani et al. [12] performed a parametric analysis of piston bowl geometry and injection nozzle configuration using three-dimensional (3D) CFD and DoE. Probst et al. [13] utilized a sequential DoE approach along with a Kriging emulator to optimize and quantify the uncertainty of a diesel engine operating point. Pei et al. [14] optimized the combustion system for a heavy-duty diesel engine operating on a gasoline-like fuel using a comprehensive CFD-DoE approach. It represented the first time that a practical engine design optimization was conducted on a supercomputer, leading to a much faster turnaround time. Park et al. [15] demonstrated an efficient method for engine calibration based on DoE which reduced the number of test measurements substantially. More recently, Pei et al. [16] performed an exhaustive CFD-DoE study to optimize a combustion system that included 256 piston bowl designs, along with their injector spray pattern, fuel injection strategy, and in-cylinder flow to achieve fuel consumption savings in a gasoline compression ignition (GCI) engine, considering multiple speed-load conditions. Khac and Zenger [17] outlined the design of static optimal control maps to attain high efficiency and emission reduction in a diesel engine. Tang et al. [18] investigated different types of piston bowl designs using CFD-DoE for a heavy-duty GCI engine across multiple load conditions. In their study, full closed-cycle engine combustion system geometry that included the valve cutouts on the piston and valve seat recessions on the cylinder fire deck was implemented in the optimization process, thereby pushing further evolvement of the state-of-the-art design optimization methodology.

Specifically, for the CFD-DoE optimization studies, significant advances have been made with the advent of more computational resources in the following aspects:
- More predictive CFD models were considered with more accurate sub-models, for example, spray, combustion, chemical kinetic, and emission models, and better mesh resolutions leading to better resolved flow fields.
- Higher numbers of design parameters were considered, from only a few parameters to about 10+ design features. This is not trivial as the sample size increases exponentially with an increase of design features.
- More realistic combustion system design features were realized by considering full cylinder, realistic engine head/piston geometries instead of sector mesh, and simplified head/piston profiles [18]. The avoidance of unrealistic design features, such as sharp edges of piston bowls, is also critical when populating the design samples [16].
- More operating conditions were considered to attain good balancing on engine performance across the full range.
- More streamlined processes were proposed with the added capability of automated geometry parameterization and mesh generations.
- Faster design optimization turnaround time was explored to shorten the engine development cycle. With practical engine combustion system designs being demonstrated on supercomputers [14,16], the CFD-DoE method has shown its advantages in drastically shortening the process [19]. Generally speaking, multiple iterations of CFD optimizations are not needed as the sample size is usually large enough to identify optimum designs [16,18].

2. Methodologies

Successful execution of a DoE optimization campaign often involves multiple iterative processes and requires a complex combination of engineering, physical, and mathematical considerations. Performing 3D CFD analysis has greatly enabled extensive design explorations. With the development of high-fidelity, predictive modeling on the physical and chemical processes, modern 3D CFD engine combustion analysis manages to adequately capture the effects of combustion system hardware design, bulk flow structure, and fuel injection strategy on engine combustion and performance. Meanwhile, the complex flow behavior often requires refined numerical mesh resolution. Also, the transient nature of the engine combustion process imposes challenges for scaling capability of CFD codes. As a result, 3D CFD optimization campaigns are often resource-limited to adopt a full factorial approach. Hence, it becomes essential to develop appropriate fractional factorial DoE methodologies to fully leverage the advantages of CFD for practical applications.

Fig. 5.1 demonstrates a conceptual process flow of CFD-DoE for engine design optimization. The rest of this section aims to elaborate on the general workflow of the CFD-DoE method. Specific emphasis is placed on the procedure and guidelines for design space construction, RSM formulation and assessment, as well as model-based design optimization and verification.

2.1 Design space construction

The high-dimensional design space highlights the complexity of engine combustion system optimization. Thus, careful construction of the design space becomes essential for CFD-DoE studies. Geometry parameters and operation parameters should be considered simultaneously in one engine design campaign, while their effects and implications could represent different features.

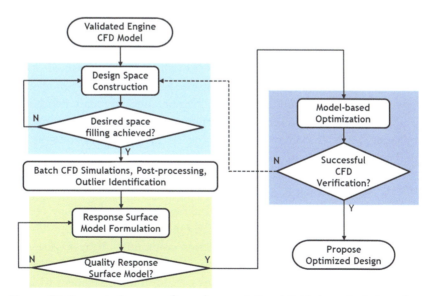

Figure 5.1 Conceptual process flow of computational fluid dynamic-design of experiments (CFD-DoE) methodology.

The shape of combustion system hardware components (piston bowl, ports, etc.) is defined by geometrical parameters. 3D engine combustion CFD combined with geometry parameterization tools offers an effective means to substantially reduce the hardware procurement and engine operation costs by exploring and evaluating various designs virtually. Piston bowl profile design of revolving piston designs is a good example. In most studies, multiple design parameters are grouped and categorized into independent and dependent variables. Independent variables often capture the key design features, for example, size of the bowl and piston step, while the dependent variables are mostly defined to ensure design constraints are conserved (e.g., compression ratio (CR), clearance height). In such a scenario, only independent geometrical parameters should be considered in the design space. Also worth noting is the necessity of conducting geometry parameterization through computer-aided design (CAD) tools. Capable CAD automation tools (e.g., CAESES [20]) streamline the geometry parameterization and surface generation for CFD simulations, thereby enabling large-scale design searches (on the order of hundreds to thousands of design variants) which could be prohibitive if using manual CAD handling.

Engine operation parameters, on the other hand, refer to the inputs that are not hardware related. Examples include fuel injection strategies such as injection timing and quantity, cam phasing, spark timing, and so on. Pure operation parameter studies are usually more efficient through experimental testing and engine calibration due to the ability for rapid generation of reliable response data. Meanwhile, the inclusion of operation parameters into CFD-DoE is warranted by the complex interactions between hardware designs and the engine operating boundary conditions. Promising performances could be obtained through either a combined approach [14,18] that involves all variables during the analysis or a discrete approach [16] that splits the two parameter groups into two consecutive optimization phases.

The range for each design variable needs to be assigned properly. The geometrical parameters defined should allow for reasonable shape change without causing discontinuity in the CAD model. In addition, they should comply with the engine mechanical design constraints. A pre-DoE evaluation is highly recommended to ensure that the ranges are appropriately defined. Sometimes large ranges are necessary for initial assessment to allow for a sufficient coverage of the design space. Subsequent iterations can then be attempted within reduced design spaces for more refined design search.

The next critical step is to fill the multidimensional hyperspace with discrete designs. The exponentially increasing sample count makes the full factorial method rarely affordable. An approach that adopts fractional factorial-based, reduced sample size is more commonly practiced. Advancements in optimization theory have suggested a few space-filling methods, among which Latin hypercube sampling [21] and the quasi-random Sobol sequences [22] have registered sizable applications. Both methods aim to distribute the samples evenly over the entire hyperspace, but with different numerical realizations. In addition, both sampling methods contain certain random components and the resulting sample sets could differ as the random seed varies. It is recommended to verify the design distributions across the design space before executing the CFD analysis. The two-dimensional (2D) and 3D plots shown in Fig. 5.2 are commonly used to inspect the sample distribution in the design space. Subsequently, adjustments can be applied to tailor the coverage of the design space.

2.2 Response surface model formulation

After completing the resource intensive CFD simulations of all sampled designs, certain standard postprocessing is typically applied to extract key performance metrics. Single value metrics (e.g., indicated specific fuel consumption (ISFC), soot, and NO_x emissions) are typically used to evaluate the design candidates. The outcome of a CFD-DoE study can then be summarized in an $M \times (N + P)$ table, where M matches the sample size, N represents the number of design variables, and P accounts for the number of performance metrics or objective variables. Further utilization of the CFD-DoE results relies on proper mathematical abstraction of the relationship between the design and objective variables.

Such desired mathematical abstraction is commonly known as a RSM, which is expected to adequately replace the expensive CFD simulations when constructed properly. Although dedicated efforts have been taken to distribute the M variants widely and evenly through the design space, it is never a trivial task to perform high-dimensional data-fit processing. Considerations regarding outlier identification and data-fit model selection are suggested for satisfactory surrogate model formulations.

In CFD-DoE studies, the response metrics are obtained from numerical simulations, which bear higher uncertainty levels than experimental measurements. Unrealistic CFD-derived responses can mislead the RSM

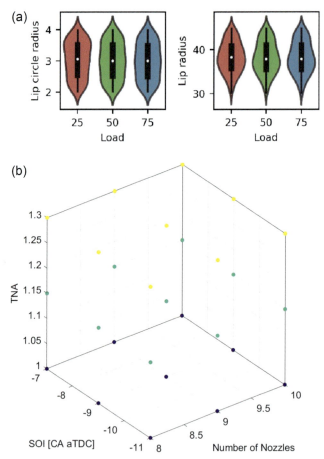

Figure 5.2 (A) Violin plots showing kernel density distribution for design variables at three engine load conditions. (B) A three-dimensional plot showing the sample distribution of design variables, start of injection (SOI), number of nozzles, and total nozzle area (TNA). *(A) Reprinted with permission from Mohan, B., Tang, M., Badra, J., Pei, Y., et al., 2021. Machine learning and response surface-based numerical optimization of the combustion system for a heavy-duty gasoline compression ignition engine. SAE Technical Paper 2021-01-0190, © SAE International. (B) Reprinted with permission from Pei, Y., Zhang, Y., Kumar, P., Traver, M., Cleary, D., Ameen, M., Som, S., Probst, D., Burton, T., Pomraning, E., et al., 2017. CFD-guided heavy duty mixing-controlled combustion system optimization with a gasoline-like fuel. SAE Int J Commerc Veh 10 (2), 532–46, © SAE International.)*

formulation and should be excluded from the result pool. Two operational guidelines can be considered for outlier identification from CFD-DoE results. First, outliers can be recognized based on statistical criteria. Extreme performance values that are too far away from the baseline require closer examination. Second, a general screening of physical outputs is always advised. Beyond those single value metrics calculated through post-processing, engine combustion characteristics, such as the cylinder pressure and heat release profiles, can provide additional diagnostic insights.

While good confidence in the CFD-DoE result pool is foundational to constructing a high-fidelity RSM, the model fitting algorithm can more dominantly affect the optimization outcome. Quite a few data-fit model options are available as open-source (e.g., DAKOTA [24]) or commercial (e.g., MATLAB MBC Toolbox [25]) software packages. The selection of the data-fit method varies between different applications and depends heavily on the complexity of the response surface of interest. Thus, the recommended practice is to evaluate multiple data-fit model options during the optimization process. A summary of popular data-fit modeling methods follows:

- Gaussian process model—also known as the Kriging method: It performs spatial interpolation between the discrete data points obtained through CFD-DoE. This method adopts a hyperparametric error model and can accommodate complicated surface forms. The resulting surrogate model returns exact replicas of the training data and can yield deteriorated predictions outside the variable ranges of the training data [26].
- Polynomial model: This model applies regression fit to the training data through liner, quadratic, or higher-order polynomial functions. Smooth responses can be expected by using this method, and it is generally believed to yield accurate approximations over small parameter spaces [27].
- Radial-basis function (RBF) method: It is not uncommon that the Kriging or polynomial models represent a reduced design domain. The RBF method provides an effective means to combine multiple local RSMs into a coherent global RSM. As such, local domain centers can be represented with enhanced fidelity [28].
- Artificial neural network (ANN) model: This method adopts the stochastic layer perceptron technique to build the ANN. Such an unconventional discipline yields unique features of the resulting response surfaces and can be useful for modeling data trends with slow discontinuities [29].

The generated RSM needs careful assessment to ensure adequate representation of the optimization subject. Similar to space filling strategy verification, statistical analysis and physical judgments should be applied to evaluate the surrogate model quality.

A k-fold cross-validation is a standard statistical tool. It first randomizes the original training dataset from CFD-DoE and partitions them into k equal sized subsamples. A single subsample is then evaluated over an intermediate surrogate model trained by the rest of the k-1 subsamples. A coefficient of prognosis (CoP) ranging from 0 to 1 can be quantified by comparing the intermediate model evaluation and a single subsample. The same process needs to be repeated for k times to yield k CoP numbers. For most data-fit models, the maximum CoP value would determine the performance of the RSM based on the general understanding that the full-dataset-trained model would outperform the best-performing intermediate model based on k-1 subsamples. One exception is the ANN method because the unconventional discipline does not necessarily favor increased sample counts. The statistical model assessment is hardly precise science. A general rule of thumb is to start with 8- to 10-fold cross-validations to pursue a satisfactory CoP of 0.9 or higher.

Physics-based model assessment may be necessary but is more involved. Therefore, 2D and 3D visualizations of multiple response curves and surfaces should be inspected. The performance trend against well-understood design variables (e.g., operation parameters such as injection timing) can be empirically judged and obvious misprediction needs to be identified. The 3D response surfaces involving operation variables are often expected to be relatively smooth based on engine performance development experiences. However, drastic geometry changes are more likely to cause complex curvatures in resulting response surfaces. Iterative processes might be necessary to adjust data-fit model selection and settings to achieve satisfactory statistical and physical behavior of the developed surrogate model.

2.3 Model-based design optimization and verification

With significant preceding efforts, an effective and low-cost RSM has been made available and ready for extensive design search. The optimization objectives should be reviewed at this stage. It is quite common to encounter multiple objective variables in practical engine optimization problems, and trade-offs often exist between these targets (e.g., NO_x vs. soot emissions). Popular multiobjective optimization algorithms specifically deal with conflicting objectives, thus screening of objective parameters becomes necessary to down-select the variables that highlight the dominant trade-offs. The excluded parameters can be assigned as "observer," over which additional constrains can be imposed. An alternative approach is to blend the multiple objective variables into a merit function and transform the task into a single-objective optimization process. Note that engineering insights into the key engine performance development targets and the priority of these targets are usually required to define meaningful merit functions.

Optimization algorithms are designed to achieve efficient searches over complex RSMs. Popular multiobjective algorithms, including the nondominated sorting-based evolutionary algorithm (NSGA-II) [30], the dominance-based multiobjective simulated annealing [31], the multiobjective genetic algorithm [24], and single-objective optimization can take advantage of the tangent search method. Meanwhile, the RSM enables a rapid design search over the design space, making it less sensitive to the efficiency of optimization algorithm. Lastly, the top design candidates generated through RSM-based optimization should be verified by executing CFD analysis to confirm the performance. Subsequently, decisions should be made on whether there is a need to iterate further a subset of the CFD-DoE workflow over an adjusted design space.

3. A recent application

Following the abovementioned methodology, a recent application is briefly presented here to illustrate this optimization process. This application optimized the closed-cycle engine combustion system of a heavy-duty diesel engine running on a market gasoline [18,23].

3.1 Engine and fuel specifications

The engine is a model year 2013 Cummins ISX15 diesel engine operating in GCI mode. The general specifications of the engine are provided in Table 5.1 and more details can be found in Zhang et al. [32]. The fuel was an antiknock index (AKI) 89 fuel, which had gasoline-like fuel properties with details reported in Tang et al. [18].

3.2 Computational fluid dynamic model setup and validation

Closed-cycle 3D CFD simulations were developed using the CONVERGE CFD solver. A summary of the models used is given in Table 5.2 and further details can be found in Refs. [16,18,35]. The CFD model was well validated against the experimental results with in-cylinder pressure and NO_x emissions comparison at various operating conditions shown in Fig. 5.3. The solid validation of CFD models provided confidence for the subsequent optimization investigation.

Table 5.1 Test engine specifications.

Displacement	14.9 L
No. of cylinders	6
Bore	137 mm
Stroke	169 mm
Compression ratio	17.3
Diesel fuel system	2500 bar common rail
Air system	Single-stage variable geometry turbocharger, High pressure cooled EGR, Charge air cooler
Engine ratings	336 kW (450 hp) @ 1800 rpm, 2373 Nm (1750 lb-ft) @ 1000 rpm

Table 5.2 Computational fluid dynamic model summary [18].

Spray models	Liquid-gas coupling: Discrete droplet modeling Injection: Blob model Collision: No-time counter model Evaporation: Frossling model Momentum exchange: Dynamic drag model Spray cone angle: Dynamic profile model [33]
Wall heat transfer model	O'Rourke and Amsden
Turbulence model	Renormalization group k-ε model
Combustion model	Detailed chemistry
Reaction mechanism	Liu et al. [34] PRF mechanism with 44 species and 139 reactions
Emission models	Hiroyasu-NSC soot model Detailed NO_x chemistry model
Fuel	AKI 89
Grid	Cartesian grid, 1.4 mm base size, 0.35 mm minimum cell size, with two levels of adaptive mesh refinement based on velocity and temperature gradients, and fixed embedding near the injector, maximum of two million cells
Time-step	Variable between 0.01 and 0.5 µs

3.3 Design variables

The design variables included several key piston bowl geometric parameters paired with three independent component-level variables: the number of injector nozzles, the spray inclusion angle, and the swirl ratio. The total

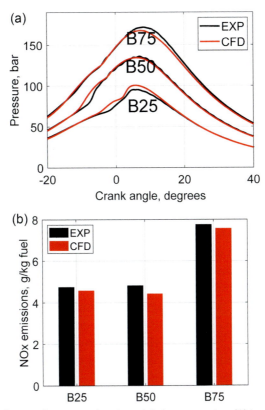

Figure 5.3 Validation of computational models by comparing (A) in-cylinder pressure and (B) NO$_x$ emissions between experiments and simulations [18].

nozzle area was conserved to keep the nozzle hydraulic flow rate constant. The spray inclusion angle was defined relative to the piston axis and the lip location. Fig. 5.4 shows an example of the parameterized bowl profile in CAESES and the definition of each geometric parameter. The parameters shown in green are independent parameters that were considered to be the primary design variables in the optimization stage. The parameters shown in red are dependent variables and were typically used to conserve the bowl volume (and thus compression ratio) and preserve a reasonable distance between the surface and critical internal structures. A total of five independent and six dependent parameters were derived from the geometry. Combining these with the three independent component-level variables, a total of eight design variables were selected as input variables in the optimization process. The Sobol sequence was used to fill the design space. Dependent variables were used by a local optimizer to finalize the geometry

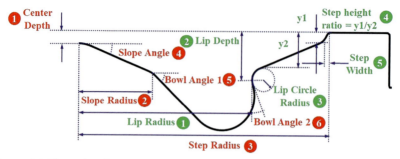

Figure 5.4 Piston bowl geometry parametrization. *(Reprinted with permission from Ref. © SAE International.)*

for each design candidate to meet the CR and other geometrical constraints. In total, 128 design candidates were generated and numerically evaluated at three engine operating conditions.

3.4 Objective variables and evaluation method

Design candidates were evaluated for their performance across the following five metrics: ISFC, soot and NO_x emissions, maximum pressure rise rate (MPRR), and peak cylinder pressure (PCP). For each design candidate, a merit value was calculated based on these five objective variables for each load using Eqs. (5.1) and (5.2).

$$\text{Merit} = 100 \left[\frac{\text{ISFC}_{\text{baseline}}}{\text{ISFC}_{\text{candidate}}} - 100 * f(\text{PCP}) - f(\text{MPRR}) - 0.1 * f(\text{Soot}) - f(\text{NO}_x) \right] \quad (5.1)$$

$$f(\text{Parameter}) = \begin{cases} 0, & \text{Parameter} \leq \text{Limit} \\ \dfrac{\text{Parameter}}{\text{Limit}} - 1, & \text{Parameter} > \text{Limit} \end{cases} \quad (5.2)$$

A weighted-average merit value was then calculated by applying specific weight from each of the different engine load conditions. Table 5.3 shows the target for each performance metric and weights for the different loads.

The merit function was designed to reward fuel consumption improvements and penalize any increase in emissions, MPRR, or PCP. The function was designed to significantly penalize an increase of PCP beyond the mechanical design limit but was less stringent on soot emissions compared to NO_x due to the intrinsic lower soot emission benefit of the GCI mode of combustion. The merit function may be designed as different forms based on the optimization targets.

Table 5.3 Performance targets and weights for individual load points.

Load	ISFC g/kWh	Soot g/kWh	NO$_x$ g/kWh	PCP bar	MPRR Bar/CAD	Weights
1375 rpm/5 bar BMEP (B25)	171.2	0.05	1	220	12	10
1375 rpm/10 bar BMEP (B50)	176.9	0.12	1			10
1375 rpm/15 bar BMEP (B75)	179.7	0.1	1.5			10

3.5 Data fitting and optimization

DAKOTA was used to construct the response surfaces in this study with the following three methods explored:

1. The Kriging method, also known as the Gaussian process spatial interpolation model, used a hyper-parametric error model that could accommodate surfaces with slope discontinuities along with multiple local minima and maxima.
2. The ANN model used a low training cost stochastic layered perceptron ANN. It is a nonparametric surface fitting method and can model data trends with slope discontinuities.
3. The second-order polynomial model (Poly2) can provide accurate approximations of the true data trends in a small portion of the parameter space.

Depending on the fitting quality, different weights were assigned to the three methods to yield weighted average performance metrics, for which a composite merit value was obtained for a specific load point. To evaluate the potential of a design under different load points, the merit values at all load points were weight averaged together to generate a single merit value.

Finally, the optimum design was obtained by using a combination of the Kriging and the second-order polynomial methods with equal weights. A single objective minimizing the composite merit value was implanted by using the NSGA-II method. After the CFD validations, the best design provided a weighted average merit value of 101.7, about 6% higher than the baseline design. The best ISFC improvement was observed as 3.2% at B25 while meeting NO$_x$ and soot emissions targets.

The best piston bowl design from the DoE optimization process (labeled as RSM) is shown in Fig. 5.5. In addition, the best design candidate selected

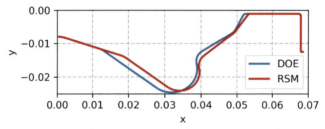

Figure 5.5 A comparison of the best designs from the design of experiments (DoE) campaign and the response surface model (RSM)-based optimization campaign. *(Reprinted with permission from Mohan, B., Tang, M., Badra, J., Pei, Y., et al., 2021. Machine learning and response surface-based numerical optimization of the combustion system for a heavy-duty gasoline compression ignition engine. SAE Technical Paper 2021-01-0190, © SAE International.)*

directly from the DoE campaign without conducting RSM-based optimization (labeled as DOE) was also provided. The baseline profile could not be plotted on the figure due to reasons of confidentiality. The values for the additional design variables are shown in Table 5.4.

4. Recommendations for best practice

To ensure the fidelity and efficiency of the CFD-DoE design optimization methodology, a few recommendations are discussed for best practice.

4.1 Adequate computational fluid dynamic model validation

To be confident of the optimization results, it is essential to have the CFD model adequately validated against engine experimental data on both combustion and emissions across multiple operating conditions. Combustion engines are extremely challenging to be accurately simulated due to the disparate length and time scales, along with a multitude of physical subprocesses (injection, evaporation, etc.) and complicated combustion

Table 5.4 Values for additional design variables.

Input parameter	Baseline	DoE	RSM
No. of nozzle holes	8	9	9
Swirl ratio	1.5	2	1.5
Spray inclusion angle	74	68.2	68

Reprinted with permission from Mohan, B., Tang, M., Badra, J., Pei, Y., et al., 2021. Machine learning and response surface-based numerical optimization of the combustion system for a heavy-duty gasoline compression ignition engine. SAE Technical Paper 2021-01-0190, © SAE International.

chemistry. Although they have progressed substantially over the years, engine CFD models still need to be sufficiently tuned and validated against experimental data.

4.2 Efficient geometry and mesh manipulation

Powerful and highly efficient geometry parameterization and mesh generation tools are critical to reduce the overhead of manual handling and warrant an efficient and streamlined optimization process.

4.3 Sample size

The sample size is an important factor in the optimization process. An appropriate balance between computational cost and the fidelity of the RSM needs to be struck. A general rule of thumb is to begin with 10 samples per design variable in the design space. The accuracy is usually acceptable and an optimum design can be obtained after a few iterations. More samples may be required for some specific applications such as cold start, where reduced robustness may occur and a larger sample size is necessary.

4.4 Optimization across full engine operation range

Fig. 5.6 shows the best designs varied across the four engine operating conditions in a heavy-GCI piston bowl geometry optimization campaign [16]. One can see that different operating conditions favored different best design profiles. A weighted merit function is generally needed to combine different operating conditions together to obtain an overall best performing design across the full engine operation range.

4.5 Computational efficiency

Traditional engine designs are usually performed with a reduced design space on computing clusters that still take months to complete. A few iterations are usually required to find the optimum designs. The effort to reduce the sample size without sacrificing accuracy and efficiency has always been the focus of exploration. On the other hand, supercomputers have excelled at their massive parallel computing capacity, usually on the order of million cores. They offer the opportunity to conduct large-scale, comprehensive design optimization campaigns. Moreover, through a large-scale design campaign, the number of iterations and turnaround time can be greatly reduced [14,16,19].

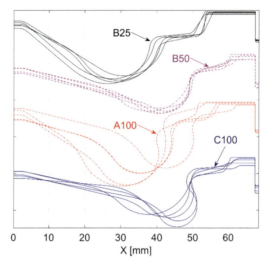

Figure 5.6 Five top performing designs at the four operating conditions. *(Reprinted with permission from Pei, Y., Pal, P., Zhang, Y., Traver, M., et al., 2019. CFD-guided combustion system optimization of a gasoline range fuel in a heavy-duty compression ignition engine using automatic piston geometry generation and a supercomputer. SAE Int J Adv Curr Pract Mobil 1 (1), 166–79. https://doi.org/10.4271/2019-01-0001, © SAE International.)*

5. Conclusions and perspectives

CFD-guided engine design optimization using DoE was introduced and discussed in this chapter. The design optimization process was presented with a specific emphasis on the procedure and guidelines for design space construction, RSM formulation and assessment, and model-based design optimization and verification. The method was demonstrated with a recent application to optimize the combustion system of a heavy-duty diesel engine running on gasoline fuel. Some insights to warrant best optimization practices are summarized as below:

- It is essential to conduct an exhaustive CFD model validation against experimental data for multiple targets at multiple operating conditions.
- Powerful and highly efficient geometry parameterization and mesh generation tools are critical to warrant an efficient and streamlined optimization.
- A sample size of 10 per design feature is a good starting point, and more samples may be required for some complicated applications.
- It is important to account for multiple operating conditions to obtain an overall best-performing design for a full range of operations.

Improving the efficiency of the optimization process has been the research focus for the industry. On one hand, it is critical to optimize sampling and optimization strategies to minimize the number of iterations to reach optimum. On the other hand, supercomputers provide a possibility of much faster turnaround time by simulating many designs concurrently. Moreover, the plentiful dataset can be utilized to create a training data for machine learning exploration. Though limited access to supercomputers is the reality, the potential for enhanced computational efficiency combined with the evolvement of the optimization approaches (e.g., machine learning) makes it a highly attractive option.

Acknowledgments

Part of the work reported in this chapter was previously published in Tang, M., Pei, Y., Guo, H., Zhang, Y., Torelli, R., Probst, D., Fütterer, C., and Traver, M., "Piston Bowl Geometry Effects on Gasoline Compression Ignition in a Heavy-Duty Diesel Engine," Journal of Energy Resources Technology, Vol 143, 062309-1, 2021.

Part of the work reported in this chapter was previously published in Mohan, B., Tang, M., Badra, J., Pei, Y. et al., "Machine Learning and Response Surface-Based Numerical Optimization of the Combustion System for a Heavy-Duty Gasoline Compression Ignition Engine," SAE Technical Paper 2021-01-0190, 2021, https://doi.org/10.4271/2021-01-0190.

The submitted manuscript was created partly by UChicago Argonne, LLC, operator of Argonne National Laboratory. Argonne, a US Department of Energy (DOE) Office of Science laboratory, is operated under Contract No. DE-AC02-06CH11357. Blues High Performance LCRC cluster facilities at Argonne National Laboratory were used for some of the simulations.

References

[1] Krottmaier J. Optimizing engineering designs. Singapore: McGraw-Hill; 1993.
[2] Besson M, Hilaire N, Lahjaily H, Gastaldi P. Diesel combustion study at full load using CFD and design of experiments. SAE Technical Paper 2003-01-1858. 2003.
[3] Han Z, Weaver C, Wooldridge S, Alger T, Hilditch J, McGee J, Westrate B, Xu Z, Yi J, Chen X, Trigui N. Development of a new light stratified-charge DISI combustion system for a family of engines with upfront CFD coupling with thermal and optical engine experiments. SAE Trans 2004;113:269—93.
[4] Mallamo F, Badami M, Millo F. Application of the design of experiments and objective functions for the optimization of multiple injection strategies for low emissions in CR diesel engines. SAE Technical Paper 2004-01-0123. 2004. https://doi.org/10.4271/2004-01-0123.
[5] Lippert AM, El Tahry SH, Huebler MS, Parrish SE, Inoue H, Noyori T, Nakama K, Abe T. Development and optimization of a small-displacement spark-ignition direct-injection engine-stratified operation. SAE Trans 2004;113:42—66.

[6] Hajireza S, Regner G, Christie A, Egert M, Mittermaier H. Application of CFD modeling in combustion bowl assessment of diesel engines using DoE methodology. SAE Technical Paper 2006-01-3330. 2006.
[7] Gothekar S, Vora K, Mutha G, Walke N, et al. Design of experiments: a systems approach to engine optimization for lower emissions. SAE Technical Paper 2007-26-012. 2007. https://doi.org/10.4271/2007-26-012.
[8] Davis R, Mandrusiak G, Landenfeld T. Development of the combustion system for general motors' 3.6L DOHC 4V V6 engine with direct injection. SAE Int J Engines 2009;1(1):85—100.
[9] Reiche D, Wooldridge S, Moilanen P, Davis G. Experimental optimization of the cold start for the EcoBoost engine. SAE Technical Paper 2009-01-1491. 2009. https://doi.org/10.4271/2009-01-1491.
[10] Catania AE, d'Ambrosio S, Finesso R, Spessa E, Cipolla G, Vassallo A. Combustion system optimization of a low compression-ratio PCCI diesel engine for light-duty application. SAE Int J Engines 2009;2(1):1314—26.
[11] Styron J, Baldwin B, Fulton B, Ives D, Ramanathan S. Ford 2011 6.7 L power Stroke® diesel engine combustion system development. SAE Technical Paper 2011-01-0415. 2011.
[12] Rajamani VK, Schoenfeld S, Dhongde A. Parametric analysis of piston bowl geometry and injection nozzle configuration using 3D CFD and DoE. SAE Technical Paper 2012-01-0700. 2012.
[13] Probst DM, Senecal PK, Chien PZ, Xu MX, Leyde BP. Optimization and uncertainty analysis of a diesel engine operating point using computational fluid dynamics. ASME J Eng Gas Turb Power 2018;140(10):102806. https://doi.org/10.1115/1.4040006.
[14] Pei Y, Zhang Y, Kumar P, Traver M, Cleary D, Ameen M, Som S, Probst D, Burton T, Pomraning E, et al. CFD-guided heavy duty mixing-controlled combustion system optimization with a gasoline-like fuel. SAE Int J Commerc Veh 2017;10(2):532—46.
[15] Park S, Kim Y, Woo S, Lee K. Optimization and calibration strategy using design of experiment for a diesel engine. Appl Therm Eng 2017;123:917—28.
[16] Pei Y, Pal P, Zhang Y, Traver M, et al. CFD-guided combustion system optimization of a gasoline range fuel in a heavy-duty compression ignition engine using automatic piston geometry generation and a supercomputer. SAE Int J Adv Curr Pract Mobil 2019;1(1):166—79. https://doi.org/10.4271/2019-01-0001.
[17] Khac HN, Zenger K. Designing optimal control maps for diesel engines for high efficiency and emission reduction. In: Proceedings of the 18th European control conference. IEEE; 2019. p. 1957—62.
[18] Tang M, Pei Y, Guo H, Zhang Y, Torelli R, Probst D, Fütterer C, Traver M. Piston bowl geometry effects on gasoline compression ignition in a heavy-duty diesel engine. J Energy Resour Technol 2021;143. 062309-1.
[19] Som S, Pei Y. HPC opens a new frontier in fuel-engine research. Comput Sci Eng 2018;20(5):77—80.
[20] Friendship Systems. CAESES user manual. 2019.
[21] McKay MD, Beckman RJ, Conover WJ. A comparison of three methods for selecting values of input variables in the analysis of output from a computer code. Technometrics 2000;42(1):55—61.
[22] Sobol IYM. On the distribution of points in a cube and the approximate evaluation of integrals. Comput Math Math Phys 1967;7(4):784—802.
[23] Mohan B, Tang M, Badra J, Pei Y, et al. Machine learning and response surface-based numerical optimization of the combustion system for a heavy-duty gasoline compression ignition engine. SAE Technical Paper 2021-01-0190. 2021.

[24] Adams BM, Bohnhoff WJ, Dalbey KR, Ebeida MS, Eddy JP, Eldred MS, Hooper RW, Hough PD, Hu KT, Jakeman JD, Khalil M, Maupin KA, Monschke JA, Ridgway EM, Rushdi AA, Seidl DT, Stephens JA, Swiler LP, Winokur JG. Dakota, a multilevel parallel object-oriented framework for design optimization, parameter estimation, uncertainty quantification, and sensitivity analysis: version 6.12 user's manual. Sandia Technical Report SAND2020-12495. November 2020.
[25] Model-based calibration Toolbox. Natick, Massachusetts: The MathWorks, Inc.; 2019. Retrieved from: https://www.mathworks.com/help/mbc/.
[26] McFarland JM. Uncertainty analysis for computer simulations through validation and calibration. PhD thesis. Nashville, Tennessee: Vanderbilt University; 2008. Available at: http://etd.library.vanderbilt.edu/ETD-db/available/etd-03282008-125137.
[27] Myers RH, Montgomery DC. Response surface methodology: process and product optimization using designed experiments. New York: John Wiley & Sons; 1995.
[28] Orr MJL. Introduction to radial basis function networks. Technical Report. Edinburgh, Scotland: University of Edinburgh; 1996.
[29] Zimmerman DC. Genetic algorithms for navigating expensive and complex design spaces. Final report for Sandia National Laboratories contract AO-7736 CA 02. 1996.
[30] Deb K, Pratap A, Agarwal S, Meyarivan TAMT. A fast and elitist multiobjective genetic algorithm: NSGA-II. IEEE Trans Evol Comput 2002;6(2):182–97.
[31] Serafini P. Simulated annealing for multi objective optimization problems. In: Multiple criteria decision making. New York: Springer; 1994. p. 283–92.
[32] Zhang Y, Voice A, Tzanetakis T, Traver M, Cleary D. An evaluation of combustion and emissions performance with low cetane naphtha fuels in a multicylinder heavy-duty diesel engine. J Eng Gas Turb Power 2016;138(10):102805.
[33] Tang M, Pei Y, Zhang Y, Tzanetakis T, Traver M, Cleary D, Quan S, Naber J, Lee SY. Development of a transient spray cone angle correlation for CFD simulations at diesel engine conditions. SAE Technical Paper 2018-01-0304. 2018.
[34] Liu YD, Jia M, Xie MZ, Pang B. Enhancement on a skeletal kinetic model for primary reference fuel oxidation by using a semidecoupling methodology. Energy Fuels 2012;26(12):7069–83.
[35] Zhang Y, Kumar P, Pei Y, Traver M, Cleary D. An experimental and computational investigation of gasoline compression ignition using conventional and higher reactivity gasolines in a multi-cylinder heavy-duty diesel engine. SAE Technical Paper 2018-01-0226. 2018.

CHAPTER 6

A machine learning-genetic algorithm approach for rapid optimization of internal combustion engines

Jihad Badra[1], Opeoluwa Owoyele[2], Pinaki Pal[2] and Sibendu Som[2]
[1]Transport Technologies Division, Research and Development Center, Saudi Aramco, Dhahran, Eastern Province, Saudi Arabia; [2]Energy Systems Division, Argonne National Laboratory, Lemont, IL, United States

1. Introduction

The global demand for energy used in the transportation sector is expected to continue rising at an annual rate of 1%–1.5% by 2040 according to recent projections [1,2]. This increase is mainly driven by the expected rise in population, gross domestic product, and living standards. Currently, internal combustion (IC) engines, fueled by petroleum-derived liquid hydrocarbons (gasoline and diesel), dominate the passenger and commercial transportation sectors with over 99% market share. IC engines are expected to remain the major source of the transportation energy demand in the interim future, despite significant growth in alternative energy and competing technologies (e.g., electric and fuel cells) [1–3].

The legislative requirements aimed at reducing tailpipe emissions, improving vehicle efficiency and mitigating the impact of transportation on CO_2 emissions are the main drivers when it comes to changes in the transportation landscape. These legislative requirements, along with consumer demands for vehicles with improved efficiency, drivability, and affordability, are pushing automakers to explore many aspects of engine design, combustion control, and after-treatment systems that simultaneously reduce fuel consumption and emissions.

IC engines have been around since the 19th century and their conceptual identity as a fuel-powered machine has not changed since. There have been significant technological improvements to their performance in response to fuel efficiency and emissions regulations [4]. The tools used to co-optimize the fuel/engine system have evolved over the years.

Until 20 years ago, experimental prototyping was the main optimizing method. It was followed by numerical simulations, including complex three-dimensional computational fluid dynamics (CFD), which played a major role in the engine/fuel system optimization. This development was enabled by the significant advancements in computing power (supercomputers, clusters, parallelization, etc.) and numerical models (turbulence, combustion, spray, heat transfer, meshing, moving boundaries, etc.). Due to high dimensionality, complexity, and highly nonlinear interactions among engine design parameters, both experimental and numerical optimization approaches can be inefficient, and take a significant amount of time and effort to obtain local rather than global optimum designs and operating conditions [5–7]. To overcome the issues with manual optimizations, alternative approaches have been developed over the years. The design of experiments (DoE) technique is commonly used for engine design optimization [8]. Genetic algorithms (GAs) [9–15] have also been used to facilitate design optimization, where an objective function is defined to represent CFD simulations. The GA approach often yields better optimum solutions compared with DoE-based optimization. Probst et al. [8] showed that the sequential DoE approach, coupled with CFD models, can successfully and efficiently optimize engines. The sequential DoE was compared to an optimization performed using a GA. Their study highlighted the strengths of both methods for optimization. The GA (known to be an efficient and effective method) found a better optimum, while the DoE method found a good optimum with fewer total simulations. The DoE method also ran more simulations concurrently, which is an advantage when sufficient computing resources are available.

Significant advancements have been made in Artificial Intelligence (AI) in the last couple of decades. These advances have enabled machine learning (ML) to be a potentially effective tool to optimize engine/fuel systems [16–19]. Particularly, neural networks have been used in several studies to efficiently optimize engine/fuel systems. The real-time engine control system correction [20,21] and the effects of fuel properties on engine emissions [7,22] are examples of some recent applications. Other ML algorithms, such as random forest and support vector machine, have also been used in engine and vehicle-related problems [23–25]. Recently, Moiz et al. [26] proposed an improved machine learning-genetic algorithm (ML-GA) approach based on an ensemble ML technique known as Super Learner [27], to optimize the operating conditions of a heavy-duty (HD) gasoline compression ignition (GCI) engine. DoE was first performed to generate a database to train an ML algorithm. The ML optimization algorithm was then used without any further CFD modeling to search for optimum engine design parameters. Their approach yielded comparable optimum operating conditions in significantly less computing time compared to the CFD-GA approach [26]. Other studies that implemented ML-GA algorithms for engine optimization may be found in the literature [7,10–12,28,29].

In this chapter, the ML-GA approach is discussed in detail. ML-GA is comprised of a robust Super Learner approach to build the surrogate model (based on simulation data), wherein multiple ML algorithms are pooled together instead of single learner. This Super Learner surroagte is then used to replace expensive simulations during the course of a GA optimization to rapidly arrive at the optimal design parameters with significantly lower cost. The repeatability of the optimization method is investigated. Moreover, automated hyperparameter optimization (HPO) strategies to enhance the efficiency and robustness of ML-GA are also discussed. In particular, a Bayesian approach is employed to optimize the hyperparameters of the base learners that make up a Super Learner model to obatin better performance. In addition to performing hyperparameter optimization (HPO), an active learning approach is leveraged, where the process of data generation, ML training, and surrogate optimization, is performed repeatedly to refine the solution in the vicinity of the predicted optimum.

2. Engine optimization problem setup

The engine considered is a four-stroke, six-cylinder Cummins ISX15 engine with a variable-geometry turbocharger, high-pressure cooled exhaust gas recirculation (EGR) loop, and charge air cooler [30]. The details of the engine configuration and baseline operating conditions are listed in Table 6.1.

The design parameters considered in the present work are listed in Table 6.2, along with their respective ranges of variation.

In total, nine input variables were chosen pertaining to fuel injector design (number of nozzle holes, total nozzle area, nozzle inclusion angle), fuel injection strategy (injection pressure, start of injection), and initial thermodynamic and flow conditions (intake valve closing temperature and pressure, EGR fraction, and swirl ratio). The ranges of variation included

Table 6.1 Engine configuration and baseline operating conditions.

Engine model	Cummins ISX15
Cylinders	6
Displacement (L)	14.9
Bore (mm)	137
Stroke (mm)	169
Connecting rod length (mm)	262
Compression ratio	17.3:1
Engine speed (rpm)	1375
Intake valve closing (IVC) (CA after top-dead-center (aTDC))	137
Exhaust valve opening (EVO) (CA aTDC)	148
Start of injection (SOI) (CA aTDC)	−9
Injection duration (CA)	15.58
Mass of fuel injected (g/cycle)	0.498
Fuel injection temperature (K)	360
Injection pressure (bar)	1600
Nozzle inclusion angle (°)	152
IVC pressure (bar)	2.15
ICV temperature (K)	323
Exhaust gas recirculation (EGR) (%)	41
Global equivalence ratio	0.57

Table 6.2 Input parameter ranges for the engine design space.

Parameter	Description	Min	Max	Unit
nNoz	Number of nozzle holes	8	10	—
TNA	Total nozzle area	1	1.3	—
Pinj	Injection pressure	1400	1800	bar
SOI	Start of injection	−11	−7	CA aTDC
Nang	Nozzle inclusion angle	145	166	deg
EGR	Exhaust gas recirculation (EGR) fraction	0.35	0.5	—
Tivc	Intake valve closing (IVC) temperature	323	373	K
Pivc	IVC pressure	2	2.3	bar
SR	Swirl ratio	−2.4	−1	—

the baseline conditions. Note that in Table 6.2, the total nozzle area is normalized with respect to its baseline value, therefore the baseline value is 1. Throughout the optimization study, the total mass of fuel injected, i.e., the engine load, was kept constant.

For optimization, an objective merit function, as shown in Eq. (6.1), was defined using indicated specific fuel consumption (ISFC, g/kW-hr) as the performance variable (to be minimized), and constraint variables based on emissions (soot, NO_x) and engine mechanical limits [peak cylinder pressure (P_{Max}), maximum pressure rise rate (MPRR)]. The merit function incurs a penalty only if soot, NO_x, P_{Max}, and MPRR exceed their constraints of 0.0268 g/kW-hr, 1.34 g/kW-hr, 220 bar, and 15 bar/CA, respectively. No penalty is incurred if these are within their prescribed limits, in which case the merit function varies only with ISFC. In addition, the constraint variables are assigned different weights to reflect their relative importance in the optimization process. A similar merit function formulation was also employed in a previous engine design optimization study [8].

$$\text{Merit normalized} = 100 \times \left[\frac{160}{\text{ISFC}} - wf1 \times f(P_{Max}) - wf2 \times f(\text{MPRR}) - wf3 \times f(\text{Soot}) - wf4 \times f(\text{NOx}) \right]$$

(6.1)

where

$$f(\text{Parameter}^*) = \begin{cases} \dfrac{\text{Parameter}}{\text{Constraint}} - 1, & \text{if Parameter} > \text{Limit} \\ 0, & \text{if Parameter} \leq \text{Limit} \end{cases}$$

(6.2)

3. Training and data examination

For an efficient optimization process, generating good quality training data is a key step toward obtaining an efficient algorithm that can model engine performance with sufficient accuracy. The quality of training data is measured by the quantity of information it conveys to a learning algorithm. For instance, training data in which the input design parameters only cover small ranges of the feasible space (training dataset 2 in Fig. 6.1) can be described as low quality because they will not teach the algorithm about the performance of the objective function in the larger feasible space. Similarly, if the training data only covers designs with a limited range of performance functions (training dataset 1 in Fig. 6.1), the algorithm would not learn the effect of each of the design parameters on the merit function. The actual behavior of the performance function is shown in black. All training

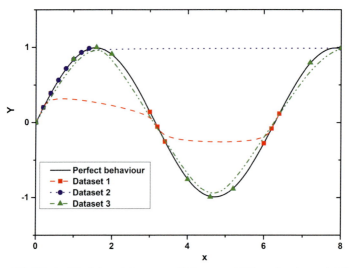

Figure 6.1 Effect of training data on predictive capabilities of polynomial fitting.

datasets contain nine data points. Training dataset 3 (green) is the most effective in obtaining good predictive capabilities as can be visually seen when comparing the perfect behavior (black line) with the predicted behaviors (lines in other colors).

Several methods can be applied to select the training data, such as Markov decision trees. These algorithms may be categorized as reinforcement learning algorithms. Their main drawback is that they must be run in a sequential manner, which limits the use of parallel computing. Other simpler methods may be employed where the relationships between input and output parameters are neglected and only the distribution of inputs is diversified as much as possible. These DoE tools are widely used and documented [8,31–33]. In this chapter, the training dataset was obtained from Moiz et al. [26] where they ran a total of 2048 simulations for the different combinations of the design variables listed in Table 6.2. The nine input variables were perturbed simultaneously within their respective ranges of variation using Monte Carlo method to generate 2048 sample (parameter) sets. The input files for the 2048 simulation cases were generated using the Converge CONGO utility [34]. The CFD simulations were run in six batches of 256 cases. The total runtime for all the simulations was around two weeks including queue time. This phase of optimization is defined as CFD-DoE in the rest of the chapter.

Before starting the optimization work, a proper data analysis from the CFD-DoE design campaign was performed to identify any outliers present within the dataset. Fig. 6.2 shows the violin plots for each of the design variables using the entire 2048-point dataset. The white dot located within each violin plot indicates the median of the data while the thick black bar in the center of the violin represents the interquartile range. The thin black lines emerging from the bar represent the rest of the distribution except for the points that were determined to be outliers using a method that was a function of the interquartile range. On each side of the black line is a kernel density estimation to show the distribution shape of the data. Wider sections of the violin plot represent a higher probability that the design candidates have this design variable value and the narrower sections represent a lower probability. The number of nozzles shows multimodal distribution with the median at the lower interquartile region. This is because the

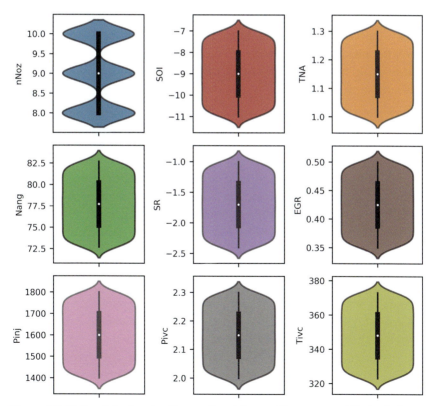

Figure 6.2 Violin plots showing the median, interquartile range, and Turkey's fences along with kernel density distribution for each of the design variables.

number of nozzles can only be an integer value between 8 and 10. The violin plot shows that nNoz is equally distributed among the three integer values. For all other variables, the median was found to be the average of the bounds given in Table 6.2. This further shows that the dataset is free of outliers, and all data points are statistically significant for the training of the models.

Many ML algorithms, such as neural networks, work better if the input data are normalized. This is because neural network parameters are usually obtained based on gradient calculations. These gradient-based problems could easily become ill-conditioned if the inputs are not normalized and have absolute values that are too different. Ill-conditioning of neural networks and the importance of data normalization have been discussed in detail by many authors [35,36]. Therefore, it is recommended to normalize the training data before feeding it to the ML algorithm. After specifying the ranges of each input parameter design, it is straightforward to normalize the input data using a min-max relation. Normalizing the output data could be tricky, especially if no obvious limits of outputs could be anticipated. In this case, the user must get as wide as possible range of outputs and normalize them accordingly, although output normalization is not as critical as input normalization.

4. Machine learning-genetic algorithm approach
4.1 Optimization methodology

In an engine optimization task, the optimization problem is first posed and different engine design parameters are defined. A sample of cases with different combinations of these parameters are solved using detailed CFD simulations or are prototyped to define the effect of these design variables on the performance of the studied engine. The performance of the engine is normally measured by a merit function that combines a certain set of output parameters which need to be defined before starting the optimization process. These sets of input variables and output parameters are utilized in a supervised ML algorithm for training purposes. The trained ML routine is then coupled with an optimization tool—in this case a GA—to search for the optimum set of engine design variables that lead to improved performance. The optimization methodology used in this chapter is shown in Fig. 6.3. To improve the predictability, an iterative process is implemented here where ML-GA was first run using the dataset consisting of 2048 points. Subsequently, the design variables and the engine output parameters

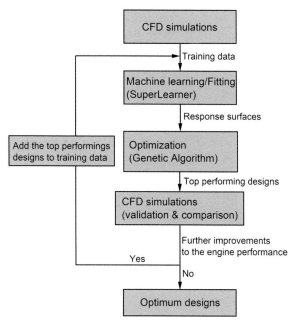

Figure 6.3 Machine learning-genetic algorithm engine optimization scheme.

of the top performing designs obtained from the optimization process are added to the training dataset. The optimum designs are then solved using CFD before adding them to the training data. This loop is repeated until no further improvements were observed after adding new designs from the previous iteration (see Fig. 6.3). This process ensures that the predictability of ML around the good designs enhances from one iteration to another.

Moiz et al. [26] used the ML package Super Learner [27], which is available as an add-on package to the statistical software R [37]. The super learning technique calculates the optimal combination of a pool of prediction algorithms, such as the arithmetic mean (SL.mean), Lasso and Elastic-Net Regularized Generalized Linear Models (SL.gmlnet), Random Forest (SL.randomForest), Support Vector Machine (SL.svm), Extreme Gradient Boosting (Xgboost), Linear regression models (SL.lm), and Feed-Forward Neural Networks and Multinomial Log-Linear Models (SL.nnet). The optimal combination is a set of these multiple models with weighting factors that minimize the cross-validated (CV) error. The GA method used to find the optimal design by Moiz et al. [26] is the GA "malschains" (memetic algorithms with local search chains) [38]. The implementation in R (Rmalschains [39]) was used in their work. Malschains uses a combination of local and global optimization techniques. The idea behind the algorithm is to apply a local search method on the most promising regions, which are found to have the highest fitness value using a (global) GA. The GA in malschains is different from a standard GA, where the individuals of the population are subjected to genetic operations simultaneously. The malschains algorithm randomly generates an initial population of individuals. The GA then evaluates the merit values (fitness) of these individuals and builds a set of individuals that can be further refined by the local search method. The main advantage of the ML-GA technique is the ability to find an optimum design for highly nonlinear problems in a short time for engine optimization tasks, where repeated CFD simulations can be time-consuming and expensive, as detailed in Moiz et al. [26]. They showed that ML-GA reduces the optimization time from about 50 days to 1 day. Despite running 2048 CFD simulations, Moiz et al. [26] showed that ML-GA can further optimize the performance of the engine using relatively low computational resources. Some of the CFD-GA and ML-GA results from Moiz et al. [26] are shown in the subsequent sections.

4.2 Repeatability of machine learning-genetic algorithm

Although ML-GA, as presented in Moiz et al. [26], presents a resource-efficient and relatively easily implemented tool for engine optimization, it can suffer from repeatability issues where the optimum set of design variables that yields the highest merit are not the same for subsequent runs. This is an anticipated behavior because the Rmalschains optimization method is based on random selections and hence running it multiple times could lead to different local solutions if not set with care. To avoid finding unrepeatable local optimum designs, special care must be taken when using certain optimization schemes. Table 6.3 summarizes the variability among the five best designs found using different parameter settings in Rmalschains, where the first line (case 0) corresponds to the base parameters used by Moiz et al. [26]. Higher than $1e^{-4}$ variance means that the merit values of the five best designs are different from each other by more than 1%. It can be clearly seen that all Rmalschains parameters affect the repeatability of the solution and hence correctly setting those parameters is critical to ensure reaching the absolute optimum design in terms of merit value. For example, the number of search iterations must be around 35,000. An iteration number of only 2400 represented by case 1 in Table 6.3 leads to five

Table 6.3 Dependence of the optimal solutions on the Rmalschains parameter settings.

Case	Popsize	Ls	I step	Effort	Alpha	Maxevals	Variance of the five best designs found
0	100	Sw	100	0.8	1	35,000	3.83E-07
1	200	Cmaes	300	0.5	0.5	2400	1.05E-01
2	100	Sw	200	0.8	1	35,000	1.63E-07
3	100	Sw	350	0.8	1	35,000	5.05E-07
4	100	Sw	3500	0.8	1	35,000	2.23E-07
5	50	Sw	3500	0.8	1	35,000	7.76E-07
6	10	Sw	3500	0.8	1	35,000	1.44E-04
7	10	Cmaes	3500	0.8	1	35,000	2.98E-07
8	10	Cmaes	3500	0.8	1	3500	7.29E-04
9	10	Cmaes	3500	0.8	1	500	7.01E-02
10	10	Cmaes	3500	0.5	1	3500	1.30E-03
11	10	Cmaes	3500	0	1	3500	7.58E-03
12	10	Cmaes	3500	0	0.5	3500	1.34E-04
13	10	Cmaes	3500	0	0.2	3500	8.22E-04
14	10	Cmaes	3500	0	0	3500	8.73E-02
15	10	Cmaes	3500	0.5	0	3500	1.88E-03
16	10	Cmaes	3500	0.5	0	3500	1.15E-01

different local optimum designs. For illustration purposes, the optimum solutions for case 1 as well as that from Moiz et al. [26] are presented in Fig. 6.4.

As seen in Fig. 6.4, the five optimal designs obtained here are different from each other despite having similar merit value (right y-axis). The observed differences can be significant for some design variables such as the number of nozzles (nNoz), start of injection (SOI), total nozzle area (TNA), nozzle angle (NozAngle), swirl ratio (SR), EGR, and injection pressure (Pinj). Only the obtained optimum intake valve closing pressure (Pivc) and temperature (Tivc) are similar for the different runs. Finding the best set of optimum solutions using Rmalschains can be a tedious exercise that requires a know-how expertise. As a remedy for this, a new method was proposed by Badra et al. [28,29] utilizing a different GA, Grid Gradient Ascent (GGA) which, although classical and forward, does not include

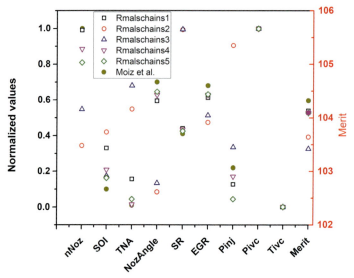

Figure 6.4 Optimum designs obtained using Rmalschains. The five designs obtained from repeating Rmalschains five times are compared with the optimum design parameters reported in Moiz et al. [26].

many setup knobs and is easy to implement with no black-boxed information. The GGA divides the multidimensional domain (d) of the design parameters into equal multidimensional cubes where the axis of each parameter is divided into n equal segments. The center of each of the n^d cubes is then used as the initial design for the optimization technique that uses the classical gradient ascent method. At the end of this step, n^d local optimum designs are obtained and the best among them is chosen as the global optimum design. Applying the ML-GGA for the case study of Moiz et al. [26] led to an optimal design that has similar merit value as the design reported by their study, as shown in Fig. 6.5. This repeatable optimum design was obtained by setting the value of n to two which requires $2^9 = 512$ iterations as compared to 35,000 iterations for the case of Rmalschains. The optimum designs from five runs are the same as the one shown in Fig. 6.5.

4.2.1 Extension of variable domain

The optimum designs reported by Moiz et al. [26] and the best designs obtained in this work have some design variables residing on or near the boundaries set by the predefined limits. These variables are the number of nozzles (nNoz), temperature at injection valve closing (Tivc), and total

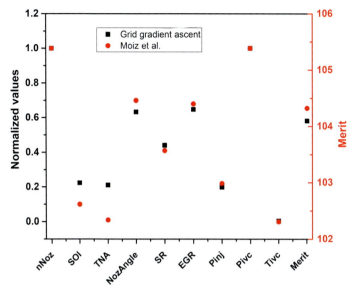

Figure 6.5 Optimum design by using machine learning-genetic algorithm versus machine learning-grid gradient ascent method. The machine learning-genetic algorithm design is taken from Moiz et al. [26].

nozzle area (TNA). Extending the design parameters range outside their initial preset limits could potentially lead to better performance. To demonstrate this, the ML-GGA algorithm using Moiz et al.'s case study is repeated by allowing 10% extension of the design parameters around their upper and lower limits. Only 10% margin is used here so that the predictability of the ML is still acceptable and within the feasible design space. The absolute values of the five best optimum designs are presented in Table 6.4 and their normalized values are shown in Fig. 6.6.

The actual outputs from CFD simulation for the five best designs obtained in this work are also included in Table 6.4. This CFD verification exercise was performed using the same CFD model employed in Moiz et al. [26]. For reference, we have also included the predicted ISFC and merit value by the ML where good predictability (within less than 0.5%) is achieved. The best ML-GA and CFD-GA results from Moiz et al. [26] are also shown in Table 6.4. As can be seen from Fig. 6.6 and Table 6.4, all optimum designs (1–5) have slightly higher merit values than that of Moiz et al. [26]. In fact, design 1 has an ISFC that is 0.14% better than the best reported by Moiz et al. [26]. Similarly, the engine-out parameters such as emissions, MPRR, and P_{Max} are better from the ML-GGA and

Table 6.4 Optimum design properties from computational fluid dynamics, reference [26] and this work.

Parameter		Design 1	Design 2	Design 3	Design 4	Design 5	ML-GA	CFD$_{MLGA}$	CFD-GA
				This work				*Reference [26]*	
nNoz		10	10	10	10	10	10	10	10
SOI		−7.98	−8.36	−8.17	−8.09	−8.08	−10.65	−10.65	−10.3
TNA		1.06	1.05	1.07	1.05	1.05	1.05	1.05	1.0
NozzleAngle		156.74	157.54	157.5	157.08	157.1	159.26	159.26	158.0
SR		−1.78	−2.02	−2.02	−1.78	−1.78	−1.81	−1.81	−1.66
EGR		0.44	0.44	0.44	0.44	0.44	0.45	0.45	0.44
P$_{inj}$		1488.70	1454.73	1496.02	1501.81	1494.11	1492.5	1492.5	1490
P$_{ivc}$		2.33	2.33	2.33	2.33	2.33	2.3	2.3	2.3
T$_{ivc}$		318.37	318.13	318.34	319.15	319.05	323.0	323.0	323.5
Engine out parameters									
Soot (g/kWh)	ML-GGA	0.0066	0.0082	0.0071	0.0069	0.0070	0.011	0.02	0.022
	CFD$_{MLGGA}$	0.0082	0.0097	0.0087	0.0084	0.0086			
NOx (g/kWh)	ML-GGA	1.33	1.32	1.34	1.33	1.33	1.32	1.23	1.28
	CFD$_{MLGGA}$	1.14	1.2	1.22	1.16	1.16			
MPRR (bar/CAD)	ML-GGA	10.32	10.30	10.67	10.36	10.30	13.28	12.22	11.31
	CFD$_{MLGGA}$	10.16	10.46	11.06	10.25	10.10			
P$_{Max}$ (bar)	ML-GGA	157.08	157.27	158.21	157.41	157.21	166.73	165.23	162.03
	CFD$_{MLGGA}$	156.78	157.18	157.90	157.20	156.97			
ISFC (g/kWh)	ML-GGA	153.16	153.18	153.18	153.21	153.21	153.37	153.97	153.85
	CFD$_{MLGGA}$	153.10	153.09	153.09	153.18	153.17			
Merit	ML-GGA	104.46	104.46	104.45	104.43	104.43	104.32	103.91	104.0
	CFD$_{MLGGA}$	104.51	104.51	104.52	104.45	104.46			

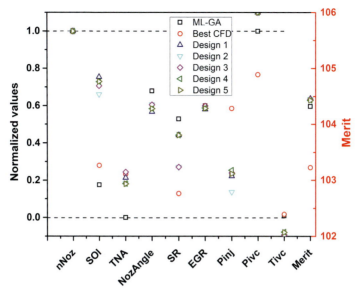

Figure 6.6 Optimum design parameters from machine learning-grid gradient ascent method with extended limits. The machine learning-genetic algorithm design and best computational fluid dynamics taken from Moiz et al. [26] are also shown.

CFD$_{MLGGA}$ compared to those reported in Moiz et al. [26]. Around 10% reductions in soot and NO$_x$ were obtained from the new simulations. The design parameters that were at their predetermined limits, such as Tivc and Pivc are now residing at the edge of the extended ranges (−0.1 and −1.1), suggesting that further extensions are potentially favorable to obtain higher merit values. However, more than 10% extension of the training data space could lead to poor prediction capability of the ML.

4.3 Postprocessing and robustness

An additional benefit of the ML-GGA technique is to take advantage of the ML algorithm for fast qualification of the optimal design in terms of robustness. In many cases, the optimum design is not unique and the optimization process produces multiple optimum designs with similar merit values. The challenge then is to further assess these results and eventually come up with a selection criterion. For engines, the main criterion that a proposed design should offer is robustness. A robust design is a design that will not rapidly deteriorate and lose its merit value once its design parameters experience small perturbations. Multiple reasons can cause perturbations in the design variables. Experimental uncertainties are the main factors for variations in the design variables. Intentional perturbations to the design variables to prevent engine-out parameters such as MPRR, P_{Max}, soot and nitric oxides (NO_x) emissions from reaching their design or regulated limits, can be another factor. To prevent the engine performance from significantly deteriorating, a robustness parameter is defined here to evaluate the sensitivity of the merit value on the design parameters. First, the permitted perturbation ranges of the design variables are defined. Subsequently, a new optimization problem can be initiated where the objective is to minimize the merit function, such that the search space is confined around the optimum design within some perturbation radius. The top five local optimum designs, reported in Table 6.4, are first saved. Afterward, a sensitivity analysis is used to evaluate the robustness of each design. The sensitivity is performed by looking for the worst design near each local optimum design. The vicinity space is defined as the multidimensional sphere centered around the local optimum design and with a radius equal to the % deterioration rate. Here, we used the Rmalschains to discover the vicinity space of each of the five best designs. The utilization of a Grid Gradient descent scheme near each of the best designs would also be possible. The comparison is reported in Fig. 6.7. Design 2 can be seen to be more robust than the other designs.

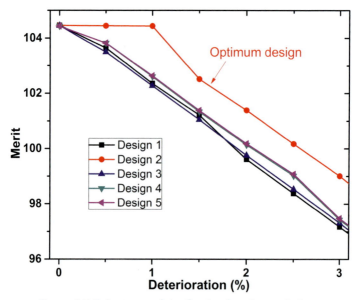

Figure 6.7 Robustness of the five local optimum designs.

5. Automated machine learning-genetic algorithm

The goal of supervised ML is typically to minimize a loss function by tuning the trainable parameters of the ML model. This usually involves calculating some measure of the error between the ML-predicted values and the actual values of the target variable. In other words, the ML model parameters are adjusted, such that the values predicted by the model closely match the desired values. In this way, supervised ML can be thought of as an optimization problem. On the other hand, many ML models contain other parameters that are not adjusted during training, known as hyperparameters. These hyperparameters are required to be initialized before training. They are used to control the learning process and their selection greatly affects the capability of the model to achieve reasonable predictions. Therefore, poorly selecting them can lead to inaccurate models. In this section, ML-GA is extended to include an intermediate step that involves selecting appropriate hyperparameters before training the ML models. The parameters and hyperparameters of different ML models considered within the Super Learner framework of ML-GA are as follows.

1. Regularized polynomial regression (RPR): The learned parameters are the weights of the transformed variables. However, the degree of the polynomial and an appropriate regularization weight, α, need to be chosen. The two hyperparameters, the regularization weight and the degree of the polynomial, control the model flexibility. Model flexibility increases with the polynomial degree, while it decreases with the aggressive regularization that comes with higher values of α.
2. *Nu* support vector regression: This involves selecting *nu*, a parameter that controls the number of support vectors, and C, a parameter that determines how much to penalize margin violations. In addition, the choice of kernel is an additional parameter. Popularly used kernels include linear, polynomial, and radial basis function kernels.
3. Kernel ridge regression (KRR): KRR learns the appropriate weight vectors that minimize the squared error loss in the chosen kernel space. The hyperparameters include the regularization strength, α, and the choice of the kernel. Depending on the choice of the kernel, there may be other hyperparameters to select. For example, the polynomial kernel requires the polynomial degree to be chosen, while the radial basis function kernel requires the selection of the gamma parameter, γ.

4. Extreme gradient boosting (XGB): As a tree-based algorithm, XGB involves learning the nodes of several regression trees. On the other hand, parameters such as the number of trees, the fraction of features and samples to use for training, the maximum tree depth, and the learning rate, need to be chosen.
5. Artificial neural networks (ANNs): Training ANNs typically involves adjusting the weights and biases via a process known as backpropagation. On the other hand, several hyperparameters need to be selected. An example is the number of hidden layers and the number of neurons in each hidden layer which both relate to the ANN architecture. Some other hyperparameters are related to the learning process itself, such as the choice of optimizer. In many cases, the optimizer too has its own hyperparameters, such as the learning rate and batch size for minibatch gradient descent, or the damping factor for the Levenberg-Marquardt [40] optimizer. The choice of activation function too is a hyperparameter, and in some cases (e.g., the slope of the Leaky ReLU [41] activation), these can also have their own hyperparameters. Other hyperparameters involve regularization weights and tolerance that control when to stop training.

5.1 Hyperparameter selection

The selection of hyperparameters can have a significant effect on the quality of the surrogate models produced. Selecting hyperparameters that confer too much flexibility to the ML models can lead to overfitting, while excessive bias may be observed when the opposite is the case. The training process of the models can also be severely hampered in some other cases, such as the selection of poor learning rates for training neural networks. Therefore, carefully selecting hyperparameters is important to obtaining accurate surrogate models.

5.1.1 Manual selection

Initial studies that introduced the ML-GA method for engine design optimization [26,28,29] relied on a combination of default hyperparameters and heuristic tuning of hyperparameters for the ML models. There are two disadvantages of these approaches. First, there is no guarantee that the default hyperparameter settings, or their manually tuned values, are optimal for a given problem. On the contrary, they often result in suboptimal ML surrogate models. Secondly, the process of manual tuning is highly subjective, time-consuming, and requires prior expert knowledge. The amount of exploration to perform around various points in the hyperparameter space is nonstandardized and depends on human judgment. Furthermore, the hyperparameters are problem-dependent and for some ML models, the number of hyperparameters to be chosen can be large. Repeatedly finding the best hyperparameters for different training tasks by manual tuning within such a high-dimensional hyperparameter space, therefore, becomes a daunting task.

5.1.2 Automated strategies for selecting hyperparameters

In contrast to a manual search technique, various methods to automate the search process have been developed. Since the goal of hyperparameter selection is to find a set of hyperparameters that lead to the best model, a way to assess the quality of a model is important. In this section, as in the previous, k-fold CV errors are used to determine the model quality in the hyperparameter selection process. The selected hyperparameters at the end of the tuning process are those that lead to the lowest CV errors. Different methods for automating the hyperparameter selection process are described below.

1. **Random search:** This involves generating random combinations of various hyperparameters by sampling from a probability distribution, and picking the hyperparameters that lead to the best model. The disadvantage of this method is that it is wasteful of computational resources in cases where evaluating the quality of different hyperparameters is compute-intensive, since it does not incorporate prior ground truth information in the search process. It also does not guarantee that good hyperparameters will be discovered within a reasonable time frame.

2. **Line or grid search:** This method involves dividing the hyperparameter space into equal segments, and sequentially evaluating the ML loss at each point. If a sufficiently fine grid is used, this method will always find good hyperparameters, since everywhere within the hyperparameter space is explored. However, the cost of performing a grid search increases exponentially with the number of hyperparameters to be found. Therefore, these kinds of methods are unsuitable for problems with a large number of hyperparameters. For approaches such as the Super Learner [42] with multiple base learners and several hyperparameters, a grid search method is impractical.

3. **Global search methods:** An alternative to the two approaches above is to cast the problem of hyperparameter selection as a separate optimization problem. In this case, the problem is treated as a black-box with the hyperparameters as the design variables and the objective being the k-fold CV loss. In this way, a global optimizer can be used to find the hyperparameters that minimize the loss function of interest. One commonly used optimizer for this process is called Bayesian optimization (BO) [43,44]. BO is a single-population adaptive surrogate method that incorporates uncertainty in guiding the hyperparameter selection process. BO works as follows:
    (a) N hyperparameters are selected and the ML model is trained using these hyperparameters to obtain an initial dataset for training the BO surrogate.
    (b) A Gaussian process regression (GPR) model that learns to predict the CV error as a function of the selected hyperparameters is trained using the dataset. The GPR model also provides an uncertainty in the predictions in form of a standard deviation, since it constructs a posterior distribution of functions.
    (c) An acquisition function, θ, is constructed as:
    $$\theta = \mu + \kappa\sigma \tag{6.3}$$
    In the above equation, μ is the mean CV error, while σ is the standard deviation obtained from the GPR model. The acquisition function in Eq. (6.3) is known as the upper confidence bound. κ controls the balance between exploration and exploitation.
    (d) A heuristic global optimizer is used to find hyperparameters that maximize the acquisition function obtained from the GPR predictions.
    (e) The ML models are trained using the set of hyperparameters from step (d) and the new CV errors are obtained.
    (f) The new data from step (e) are added to the existing dataset.
    (g) The algorithm returns to step (b) to complete the loop. The loop is continued until convergence.

As a global optimization approach, BO incorporates exploration and exploitation into the search process. In the acquisition function, exploration comes from the standard deviation, while exploitation is provided by the mean predictions. Exploration is important, since it makes the optimizer examine the hyperparameter space extensively without leaving out any region that may contain promising hyperparameters. On the other hand, exploitation helps the model search areas of the hyperparameter space that appear promising. The balance between these two is controlled by the value of κ. Small values of κ lead to the exploitation term being more dominant, and vice-versa. While lower values of κ may lead to quick convergence, it also increases the risk of converging to a local optimum prematurely. On the other hand, too much exploration, while avoiding convergence to a local optimum, may unnecessarily prolong time to converge. In this section, the value of κ was set at 2.576.

5.2 Problem setup

The problem considered in this section is the same as in the "Engine optimization problem setup" section, which involves the optimization of a compression-ignition engine operating on a gasoline-like fuel. The operating conditions, the merit function, and variable bounds are shown in Table 6.1, Eq. (6.1), and Table 6.2, respectively. The automated selection of hyperparameters for design optimization is referred to as automated machine learning-genetic algorithm (AutoML-GA) [45,46] in this chapter, in contrast to ML-GA which involves the use of default hyperparameters. It must be noted that the optimization performed here is different from the study by Moiz et al. [26] in two aspects: (1) It uses much smaller datasets to train the Super Learner and uses an active learning strategy, and (2) the number of active learning iterations is fixed to 20.

The optimization process starts by generating initial data points, Np using Latin hypercube sampling with multi-dimensional uniformity. In this section, results are based on $Np = 150$. At these points, CFD simulations are run to determine their respective design merits. A database is constructed, which contains the design variables and corresponding merits of these data points. Before training the Super Learner surrogate, hyperparameter selection using BO is performed to determine the appropriate hyperparameters that achieve low CV errors. In this chapter, different hyperparameters were used for the prediction of ISFC, soot, NO_x, P_{Max}, and MPRR. The hyperparameter selection process was limited to the training process of the base learners, as opposed to the way these base learners are combined in the Super Learner. For a given target variable (e.g., ISFC), in order to obtain the Super Learner prediction, HPO was performed for all the base learners independently. In this way, the hyperparameters that lead to the lowest k-fold CV errors in the prediction of that specific target variable were obtained for each base learner. These base learners were then combined using appropriate weights that minimize the CV error. As in the "ML-GA approach" section, active learning was employed to progressively refine the solution near the global optimum obtained based on the initial training. The hyperparameter selection process was only performed before the beginning of the active learning loop, as shown in Fig. 6.8. For subsequent active learning iterations, the Super Learner was updated by retraining the base learners and recombining them to reflect the new ground truth information obtained from CFD. To speed up the active learning process, five global optima were added to the database at each iteration, obtained by running the five GA optimization processes independently to find the global optimum based on the Super Learner's predictions. The AutoML-GA approach used in this chapter is illustrated in Fig. 6.8. Table 6.5 summarizes the list of hyperparameters, and the bounds within which they were varied. In some cases, where the hyperparameters were varied over a large range, the search was performed on a log scale. The default hyperparameter values are based on the default values in Scikit-learn version 0.22.2 [47].

5.3 Results

In this section, AutoML-GA is compared to the use of default values for the hyperparameters (ML-GA). Fig. 6.9 shows scatter plots of the actual values of various target variables against the values obtained from ML-GA (left) and AutoML-GA (right). In the case where the ML prediction is exactly equal to the CFD solution, all the scatter points will lie exactly on the diagonal. In reality, prediction errors are inevitable; hence, the figures contain some deviation from the ideal diagonal lines. The figures also contain information regarding the R^2 values, which measure the degree of bias in the trained models. Higher values of R^2 are desirable, since these mean that the ML predictions are closer to the ground truth solutions. From the figure, it can be observed that AutoML-GA leads to much better agreements with the CFD data. The plots contain significantly less scatter with higher R^2 values, compared to ML-GA. Since the overall goal of surrogate optimization is to find the design optimum using the trained ML models, it is important that these surrogates closely represent the actual CFD solution. Excessive deviation from the ground truth can lead to the predicted optimum being far from the true optimum, as will be shown later in this chapter.

A machine learning-genetic algorithm approach

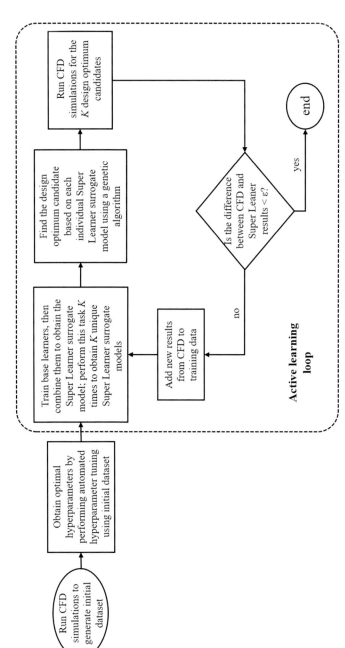

Figure 6.8 Schematic illustrating AutoML-GA workflow.

Table 6.5 List of hyperparameters and their bounds.

Base learner	Parameter	Min value	Max value	Log scale
RPR	Regularization weight, α	-6	0	Yes
RPR	Polynomial degree	1	10	No
SVR	Parameter to control the number of support vectors, v	1×10^{-10}	1.0	No
SVR	Penalty parameter, C	-6	2.5	Yes
SVR	Kernel coefficient, γ	-6	0	Yes
KRR	Regularization parameter, α	-6	0	Yes
KRR	Gamma parameter, γ	-4	0	Yes
XGB	Number of trees	2	4	Yes
XGB	Learning rate	-3	0	Yes
XGB	Maximum tree depth	2	8	No
ANN	Number of neurons in each hidden layer	10	250	No
ANN	L2 regularization parameter, α	-6	0	Yes
ANN	Tolerance to determine if training should be stopped early, tol	-6	-2	Yes

Each optimizer was tested for a total of 10 independent trials, each one performed by reinitializing the problem and repeating the optimization process. Some base learners such as neural networks and tree-based algorithms contain nondeterministic elements, making each trial different. Fig. 6.10 shows the average mean-squared error (AMSE) that occurs in approximating various quantities for different base learners. The average is calculated across all the trials performed, based on the initial training dataset before the active learning loop begins. The figures show, as expected, that the Super Learner leads to smaller errors than the best-performing base learner. In addition, AutoML-GA achieves lower prediction errors compared to ML-GA for all base learners. Overall, AutoML-GA reduces the Super Learner's AMSE obtained using ML-GA by 26%, 40%, 33%, and 34% in predictions of ISFC, soot, NO_x, and MPRR, respectively.

The optimum designs obtained using ML-GA and AutoML-GA are compared with the baseline design in Table 6.6. From the table, it can be seen that both ML-GA and AutoML-GA succeed in finding designs for which the values of soot, NO_x, MPRR, and P_{Max} do not exceed the constraints defined in Eq. (6.2). From the table, it can be seen that the ISFC

A machine learning-genetic algorithm approach   149

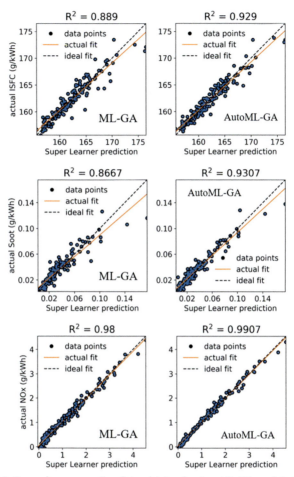

Figure 6.9 Super learner parity plots obtained using ML-GA and AutoML-GA.

of the baseline, ML-GA, and AutoML-GA are 156.53, 154.13, and 153.69 g/kWh, respectively. This corresponds to a reduction in fuel consumption of 1.81% for AutoML-GA, compared to 1.53% for ML-GA. The pressure traces (shown in Fig. 6.11) are consistent with these findings. AutoML-GA and ML-GA have significantly higher in-cylinder pressures, compared to the baseline. Since the injected fuel mass is constant, the optimum designs obtained using ML-GA and AutoML-GA would be expected to produce more work per cycle, which can be visually deduced from the areas under the in-cylinder pressure traces in the figure.

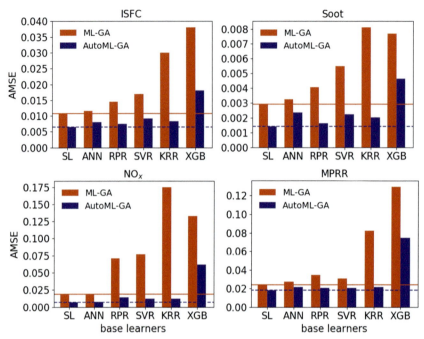

Figure 6.10 AMSE of ML-GA and AutoML-GA for various base learners and target variables.

The evolution of ISFC and NO$_x$ with the number of active learning iterations for AutoML-GA is shown in Fig. 6.12 as a population-based plot. The individual symbols for each iteration correspond to the five designs run per iteration (as shown in Fig. 6.8). In the beginning, ISFC has a high value (poor fuel economy) initially, but then decreases as the optimization progresses. At the same time, the amount of NO$_x$ emissions increases but flattens as the threshold is approached. The designs containing points that violate the NO$_x$ constraints (depicted with red circles) are penalized. The best design (depicted by the blue circle) attained by AutoML-GA produces NO$_x$ emissions that remain below the threshold while simultaneously achieving a low ISFC.

Contour plots of equivalence ratio and temperature are shown in Fig. 6.13 at 15 crank angle degrees after top dead center (aTDC) along a vertical cut plane. In the plots, comparisons of the baseline, ML-GA, and AutoML-GA are displayed. The temperature plots (left) show that ML-GA and AutoML-GA have higher peak temperatures, compared to the

Table 6.6 Baseline design and best designs obtained from ML-GA and AutoML-GA.

Design variables			
	Baseline	ML-GA	AutoML-GA
nNoz	9	10	10
SOI (CA aTDC)	−9.00	−9.00	−10.09
TNA	1.00	1.02	1.00
NozzleAngle (deg)	152.00	155.02	158.31
SR	−1.00	−2.40	−1.68
EGR	0.41	0.44	0.44
P_{inj} (bar)	1600.00	1794.15	1416.41
P_{inv} (bar)	215,000	229,623	229,688
T_{ivc} (K)	323.00	324.33	324.32

Output variables			
	Baseline	ML-GA	AutoML-GA
ISFC (g/kWh)	156.53	154.13	153.69
Soot (g/kWh)	0.0235	0.010345	0.014007
NOx (g/kWh)	1.07	1.32	1.31
MPRR (bar/CA)	11.22	14.07	10.86
P_{Max} (bar)	152.31	161.85	159.93
Merit	102.2	103.81	104.10

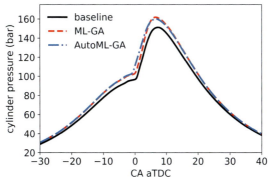

Figure 6.11 Comparison of the in-cylinder pressure trace for the baseline, ML-GA's best case, and AutoML-GA's best case.

baseline. This observation is consistent with the lower soot, and slightly higher NO_x produced by their respective optimum designs in Table 6.6. The equivalence ratio plots (left) show that the baseline case leads to larger pockets of rich fuel-air mixtures, as highlighted by the red dashed lines in

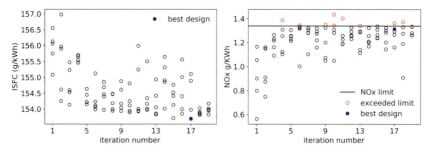

Figure 6.12 Population-based plots showing the evolution of indicated specific fuel consumption (left) and NO_x (right) with the number of active learning iterations for AutoML-GA.

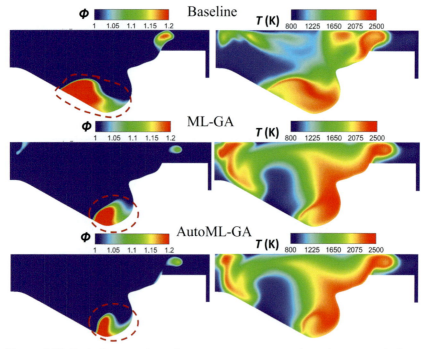

Figure 6.13 Equivalence ratio and temperature contour plots along a vertical cut plane for the Baseline case (top), ML-GA best case (middle), and AutoML-GA best case (bottom). Red dashed region highlights the predominant fuel-rich mixtures.

the plots. This is attributed to better fuel-air mixing for the ML-GA and AutoML-GA best designs owing to a larger number of injector nozzle holes and higher swirl ratios.

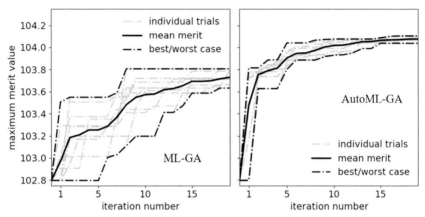

Figure 6.14 Maximum merit values as a function of the number of active learning loops using ML-GA (left) and AutoML-GA (right).

We now shift our attention to how efficient ML-GA and AutoML-GA are in finding the design optimum. Fig. 6.14 shows the best merits obtained from ML-GA and AutoML-GA as functions of the number of active learning loops completed. In the figure, the solid lines represent the mean performance over the 10 trials, while the dashed lines represent the best and worst performances. The gray dashed lines represent individual trials. Both optimizers start from a merit value of 102.8 at iteration 0, which corresponds to the best merit obtained from the Latin hypercube sampling over the design space. It can be seen from the figure that after the first active learning loop, AutoML-GA is able to reach a higher merit of 103.5 on average, compared to 103.2 for ML-GA. This is expected, since as shown in Figs. 6.9 and 6.10, AutoML-GA leads to a better approximation of the actual CFD solution. As the active learning loop progresses, AutoML-GA consistently maintains superior merit values, compared to ML-GA. The average merit value after 20 iterations is 104.08 for AutoML-GA, yielding an improvement of 1.2% over the initial merit, while for ML-GA leads to an average merit of 103.73 which corresponds to an improvement of 0.9% after 20 iterations.

In this case, the active learning loop was assumed to be converged once the absolute difference between the predicted merits and the actual merits at the newly sampled points, ε, falls below 5×10^{-3} without improvements to the merit for four iterations. The values of ε obtained from ML-GA (left)

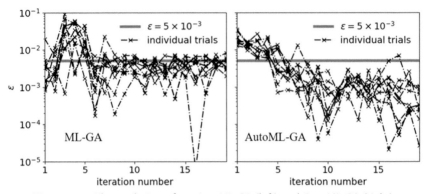

Figure 6.15 The evolution of ε using ML-GA (left) and AutoML-GA (right).

and AutoML-GA (right) are shown in Fig. 6.15. The black lines depict the individual trials. Here, it can be seen that for AutoML-GA, ε starts at a high value and gradually decreases as more active learning loops are completed until it falls below 5×10^{-3}. This means that the optimized Super Learner surrogate leads to predictions which are close to the actual CFD solution. In contrast, ML-GA does not see a similar downward trend in the value of ε; the values remain high throughout the 20 iterations. This suggests that even after many iterations, ML-GA produces significant errors in its approximation of the CFD solution.

Finally, a summary of the optimized quantities compared to the default values is shown in Fig. 6.16 for a subset of the full list of hyperparameters from Table 6.5. It can be seen that in many cases, the optimal hyperparameters are far from the default values. An example of this is the number of trees for XGB, where the optimal hyperparameter values found for various target variables are significantly higher than the default value of 1000. This observation, coupled with the shallower optimized trees (as evident from the plot showing the maximum tree depth), suggests that the default hyperparameters are prone to overfitting the data. Another observation is that various target variables may require different optimal hyperparameters. For example, the optimal kernel coefficient for SVR varies significantly between the target variables, from a high of 0.2 for ISFC to a low of 0.05 for P_{Max}. This further demonstrates the importance of automated methods for selecting hyperparameters because manual selection of multiple hyperparameters via trial and error for multiple target variables can be a time-consuming process.

A machine learning-genetic algorithm approach 155

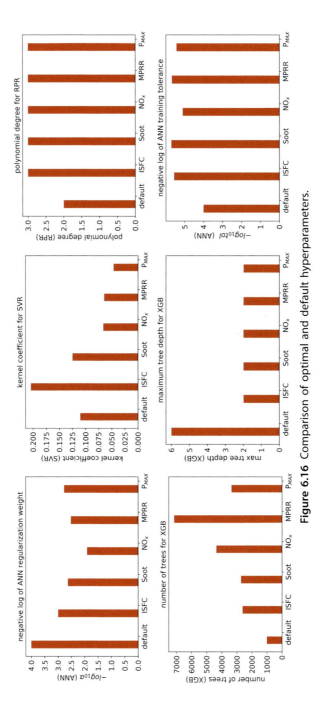

Figure 6.16 Comparison of optimal and default hyperparameters.

6. Summary

In this chapter, a methodical approach for tackling engine design optimization problems using a ML-GA was presented. Critical precautions, recommended algorithms, and suitable optimization techniques were discussed. Additional analyses and improvements were introduced to the ML-GA code, and its potential was demonstrated using an engine optimization case study focused on optimizing the operating conditions of a HD engine operating in GCI combustion mode. It was shown that active learning, automated HPO, and choice of GA optimizer can further lend robustness and sample-efficiency to ML-GA.

Acknowledgments

Part of the work reported in this chapter was previously published in Badra, J. A., Khaled, F., Tang, M., Pei, Y., Kodavasal, J., Pal, P., Owoyele, O., Fuetterer, C., Mattia, B., and Aamir, F. "Engine Combustion System Optimization Using Computational Fluid Dynamics and Machine Learning: A Methodological Approach." ASME. Journal of Energy Resources Technology, 2021, 143(2), https://doi.org/10.1115/1.4047978.

This work has been supported by the Transport Technologies Division at Saudi Aramco R&DC. We would also like to thank Aramco Americas for their support with the computing cluster at the Aramco Research Center Houston.

The submitted manuscript was created partly by UChicago Argonne, LLC, Operator of Argonne National Laboratory. Argonne, a US Department of Energy (DOE) Office of Science laboratory, is operated under Contract No. DE-AC02-06CH11357. This research was partly funded by the US DOE Office of Vehicle Technologies, Office of Energy Efficiency and Renewable Energy under Contract No. DE-AC02-06CH11357. Blues High Performance LCRC cluster at Argonne National Laboratory was used for the simulations.

References

[1] ExxonMobil, outlook for energy: a view to 2040. Texas: ExxonMobil; 2018.
[2] (EIA), U.S.E.I.A., international energy outlook 2018. 2018 [Washington, DC].
[3] Kalghatgi GT. The outlook for fuels for internal combustion engines. Int J Engine Res 2014;15(4):383–98.
[4] Kesgin U. Genetic algorithm and artificial neural network for engine optimisation of efficiency and NOx emission. Fuel 2004;83(7):885–95.
[5] Gen M, Cheng R. Genetic algorithms and engineering optimization, vol. 7. John Wiley & Sons; 2000.
[6] Manolas DA, et al. Operation optimization of an industrial cogeneration system by a genetic algorithm. Energy Convers Manag 1997;38(15–17):1625–36.
[7] Wong KI, et al. Modeling and optimization of biodiesel engine performance using advanced machine learning methods. Energy 2013;55:519–28.
[8] Probst DM, et al. Optimization and uncertainty analysis of a diesel engine operating point using computational fluid dynamics. J Eng Gas Turbines Power 2018;140(10):102806.

[9] Zhang Q, Ogren RM, Kong S-C. A comparative study of biodiesel engine performance optimization using enhanced hybrid PSO—GA and basic GA. Appl Energy 2016;165:676—84.
[10] Wickman DD, Senecal PK, Reitz RD. Diesel engine combustion chamber geometry optimization using genetic algorithms and multi-dimensional spray and combustion modeling. SAE Int 2001;110:487—507.
[11] Hanson R, et al. Piston bowl optimization for RCCI combustion in a light-duty multi-cylinder engine. SAE Int J Engines 2012;5(2):286—99.
[12] Bertram AM, Zhang Q, Kong S-C. A novel particle swarm and genetic algorithm hybrid method for diesel engine performance optimization. Int J Engine Res 2016;17(7):732—47.
[13] Shi Y, Reitz RD. Optimization of a heavy-duty compression—ignition engine fueled with diesel and gasoline-like fuels. Fuel 2010;89(11):3416—30.
[14] Wu Z, Rutland CJ, Han Z. Numerical optimization of natural gas and diesel dual-fuel combustion for a heavy-duty engine operated at a medium load. Int J Engine Res 2018;19(6):682—96.
[15] Broatch A, et al. Numerical methodology for optimization of compression-ignited engines considering combustion noise control. SAE Int J Engines 2018;11(6):625—42. https://doi.org/10.4271/2018-01-0193.
[16] He Y, Rutland CJ. Modeling of a turbocharged di diesel engine using artificial neural networks. SAE Tech Pap 2002;111:1532—43.
[17] He Y, Rutland CJ. Neural cylinder model and its transient results. SAE Tech Pap 2003. 2003-01-3232.
[18] Samadani E, et al. A method for pre-calibration of DI diesel engine emissions and performance using neural network and multi-objective genetic algorithm. Iran J Chem Chem Eng (Int Engl Ed) 2009;28(4):61—70.
[19] Vaughan A, Bohac SV. A cycle-to-cycle method to predict HCCI combustion phasing. In: ASME 2013 internal combustion engine division fall technical conference. American Society of Mechanical Engineers; 2013.
[20] Krijnsen HC, et al. Optimum NOx abatement in diesel exhaust using inferential feedforward reductant control. Fuel 2001;80(7):1001—8.
[21] Malikopoulos AA, Assanis DN, Papalambros PY. Real-time self-learning optimization of diesel engine calibration. J Eng Gas Turbines Power 2009;131(2):022803.
[22] de Lucas A, et al. Modeling diesel particulate emissions with neural networks. Fuel 2001;80(4):539—48.
[23] Orfila O, Saint Pierre G, Messias M. An android based ecodriving assistance system to improve safety and efficiency of internal combustion engine passenger cars. Transport Res C Emerg Technol 2015;58:772—82.
[24] Bergmeir P, et al. Using balanced random forests on load spectrum data for classifying component failures of a hybrid electric vehicle fleet. In: 2014 13th international conference on machine learning and applications; 2014.
[25] Rychetsky M, Ortmann S, Glesner M. Support vector approaches for engine knock detection. In: IJCNN'99. International joint conference on neural networks. Proceedings (Cat. No.99CH36339); 1999.
[26] Moiz A, Pal P, et al. A machine learning-genetic algorithm (ML-GA) approach for rapid optimization using high-performance computing. SAE Int J Commer Veh 2018;11(5):291—306. https://doi.org/10.4271/2018-01-0190.
[27] Polley EC, Van Der Laan MJ. Super learner in prediction. 2010.
[28] Badra J, et al. Combustion system optimization of a light-duty GCI engine using CFD and machine learning. SAE Int 2020. 2020-01-1313.

[29] Badra JA, et al. Engine combustion system optimization using computational fluid dynamics and machine learning: a methodological approach. J Energy Resour Technol 2020;143(2):022306. https://doi.org/10.1115/1.4047978. JERT-20-1594.
[30] Pei Y, et al. CFD-guided heavy duty mixing-controlled combustion system optimization with a gasoline-like fuel. SAE Int J Commer Veh 2017;10(2):532−46.
[31] Fritz S, Hötzendorfer H, Koller M. Design of experiments in large diesel engine optimisation. MTZ Ind 2014;4(1):40−5.
[32] Wilson VH. Optimization of diesel engine parameters using Taguchi method and design of evolution. J Braz Soc Mech Sci Eng 2012;34(4):423−8.
[33] Hicks CR, Turner KV. Fundamental concepts in the design of experiments, vol. 40. New York: Oxford University Press; 1999.
[34] Senecal P, Richards K, Pomraning E. CONVERGE (version 2.4.0) manual. Madison, WI: Convergent Science Inc.; 2018 (2014).
[35] Sola J, Sevilla J. Importance of input data normalization for the application of neural networks to complex industrial problems. IEEE Trans Nucl Sci 1997;44(3):1464−8.
[36] Saarinen S, Bramley R, Cybenko G. Ill-conditioning in neural network training problems. SIAM J Sci Comput 1993;14(3):693−714.
[37] Polley E, et al. Package 'SuperLearner'. CRAN; 2018.
[38] Bergmeir CN, Molina Cabrera D, Benítez Sánchez JM. Memetic algorithms with local search chains in R: the Rmalschains package. American Statistical Association; 2016.
[39] Bergmeir C, Molina D, Benıtez J. Rmalschains: continuous optimization using memetic algorithms with local search chains (MA-LS-Chains) in R. J Stat Software 2012.
[40] Levenberg K. A method for the solution of certain non-linear problems in least squares. Q Appl Math 1944;2(2):164−8.
[41] Maas AL, Hannun AY, Ng AY. Rectifier nonlinearities improve neural network acoustic models. In: Proc. ICML.; 2013.
[42] Van der Laan MJ, Polley EC, Hubbard AE. Super learner. Stat Appl Genet Mol Biol 2007;6(1).
[43] Močkus J. On Bayesian methods for seeking the extremum. In: Optimization techniques IFIP technical conference. Springer; 1975.
[44] Mockus J. Bayesian approach to global optimization: theory and applications, vol. 37. Springer Science & Business Media; 2012.
[45] Owoyele O, Pal P. A Novel Active Optimization Approach for Rapid and Efficient Design Space Exploration Using Ensemble Machine Learning. J Energy Resour Technol 2020;143(3):032307. https://doi.org/10.1115/1.4049178. JERT-20-1642.
[46] Owoyele O, Pal P, et al. Application of an automated machine learning-genetic algorithm (AutoML-GA) coupled with computational fluid dynamics simulations for rapid engine design optimization. Int J Engine Res 2021. https://doi.org/10.1177/14680874211023466.
[47] Pedregosa F, et al. Scikit-learn: machine learning in Python. J Mach Learning Res 2011;12(October):2825−30.

CHAPTER 7

Machine learning—driven sequential optimization using dynamic exploration and exploitation

Opeoluwa Owoyele and Pinaki Pal

Energy Systems Division, Argonne National Laboratory, Lemont, IL, United States

1. Introduction

Design optimization, which involves the selection of the best design among finite or infinite alternatives, is a key element in the design process of internal combustion (IC) engines. In contrast to the traditional design process that involves the use of experiments, simulation-driven design optimization (SDDO) involves the use of computational fluid dynamics (CFD) simulations to drive optimization. In this regard, optimizers that use predefined rules to tune designs, coupled with CFD simulations can play a vital role. Simulations allow engineers to virtually test multiple engine designs within large design spaces without incurring costs associated with experimental prototyping. Some parameters that are difficult to tune using experiments (e.g., piston bowl shape, number of injector holes) can also be readily adjusted and studied using simulations. Therefore, the use of simulations, coupled with design optimizers, is a powerful tool that can be used to explore design spaces extensively in an automated fashion and aid the development of cleaner and more efficient engines.

However, a major drawback of SDDO lies in the compute-intensive nature of CFD simulations. Developing computational models for IC engines often involves capturing multiphysics phenomena involving turbulent gas dynamics, spray injection and breakup, combustion, heat transfer, etc. Capturing all these phenomena and the interactions among them is challenging. While a lot of work in recent decades has focused on developing tractable modeling approaches for IC engines, a lot of work remains to be done in the area of simulation-efficient design optimizers that can utilize these models to drive design decisions. In this regard, of utmost importance

Artificial Intelligence and Data Driven Optimization of Internal Combustion Engines
ISBN 978-0-323-88457-0
https://doi.org/10.1016/B978-0-323-88457-0.00001-1

Copyright © 2022 Elsevier Inc. All rights reserved. UChicago Argonne, LLC,
Contract No: DE-AC02-06CH11357.

159

is the development of design optimizers that are robust enough to find optimum designs with the use of limited simulation data. This is important since, even with the use of reduced models, CFD simulations can take several hours, days, or weeks, to complete. As such, optimizers that require several simulations in parallel over hundreds of successive design iterations are not suitable for IC engine design optimization.

In recent years, machine learning (ML) has emerged as a promising tool to develop design optimization schemes that reduce either the overall design time, or the number of simulations needed, or both [1–5]. Typically, these ML models take the design variables as inputs and return the design objective as output. In this way, the ML models are trained on CFD data to provide a fast approximation for the CFD simulation. However, the amount of data required to build these ML models is typically unknown a *priori*. Furthermore, in many cases, the need for hundreds of data points to train the ML models may require more computational resources than are available. Therefore, in this chapter, a active ML–embedded optimization (ActivO) approach is presented [6,7]. ActivO addresses the issues with existing methods for design optimization [5,8,9] by markedly reducing the number of CFD simulations required to achieve convergence to the optimum design. Besides, the data generation and ML training are tightly coupled: the ML models are progressively updated as more data are realized from CFD simulations, and these ML models in turn guide the data generation process to ensure rapid convergence to the global optimum. The details of ActivO algorithm are presented in the next section.

2. Active ML optimization (ActivO)

2.1 Basic algorithm

The ActivO algorithm incorporates two ML models—a strong learner and a weak learner, both coupled to the CFD solver. Using data from CFD simulations, both learners learn to predict the design objective as a function of the design variables. The strong learner is a low-bias, high-variance learner, specifically designed to overfit the data. On the other hand, the weak learner is designed to underfit the data, by providing a high variance. An illustration of how the two learners broadly differ from each other is shown in Fig. 7.1 below. The figure shows that while the strong learner creates a fit that passes through all the available ground truth data points, the weak learner only attempts to capture the general trend of the data.

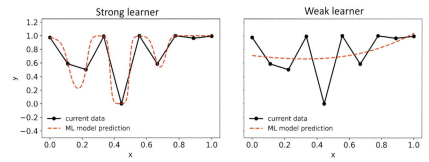

Figure 7.1 Illustration of fits produced by the strong learner (left) and weak learner (right).

The active ML optimization process works as follows:
1. Run CFD simulations at N points in the design space.
2. Post-process the CFD results, extract objective (merit) values, and add new data to the database.
3. Train weak and strong learners on the updated database from step 2.
4. Determine the next set of N points to run. If converged, terminate optimization. Else, return to step 1.

The process described above is an active ML process. Here, the ML models are not passive in the data generation process. Rather, they guide the selection of points from the design space to simulate in the next iteration. Active learning methods have various query strategies to determine the next points for which data should be generated. In the basic ActivO algorithm, simulations for N unique design points are run in each design iteration. The active learning strategy is a hybrid one, where some of the next design points are based on the query strategy of the weak learner, and the others are based on the query strategy of the strong learner. In the basic ActivO algorithm, for N design points per design iteration, only one query point comes from the strong learner, while the selection of the other $N-1$ points is based on the weak learner.

2.2 Query strategies

The two important elements of performing sequential design optimization, exploration and exploitation, are incorporated in the operation of ActivO. Exploration involves querying uncharted regions of the design space. This is important to avoid premature convergence to local optima. Exploration comes from the weak learner, where points that have weak learner-predicted merit values that are better than some (user-specified) threshold value are selected for evaluation in the next iteration. The weak learner, being a low variance learner, provides an approximate idea of what regions of the design space are likely to contain the design optimum. By randomly sampling within these promising regions as predicted by the weak learner, needlessly running simulations in poor regions of the design space is avoided. In this work, support vector regression [10] is used as the weak learner. Other options are to use low-order polynomial regression models or kernel ridge regression. At each design iteration, a large number of nominees are randomly generated over the design space, and the weak learner is used to evaluate the fitness values at these points. A two-stage selection process is used to arrive at the query points for the next iteration. First, points with low fitness values as predicted by the weak learner are eliminated, as it is assumed that these points are in nonpromising regions of the design space. To do this, the points are ranked based on the predicted fitness values, and the fitness value corresponding to the kth percentile, λ_k, is determined. Nominees that rank below this percentile are discarded. In the second stage, the nominees that are farthest from the already sampled points are chosen. This is done by calculating the minimum distance between each remaining nominee and all the already sampled points, d_{min}. The nominee with the highest value of d_{min} is selected. This two-stage selection process is repeated until the desired number of weak learner query points are obtained. A schematic illustrating the process described above is shown in Fig. 7.2.

On the other hand, the strong learner enables exploitation, which involves searching around the best-known optimum to find even better designs. This is done by using a global optimizer (e.g., a genetic algorithm or swarm-based optimizer) to find the optimum point based on the strong learner prediction. As more simulation data are generated over multiple iterations, the strong learner becomes a more accurate surrogate model close to the best-known optimum, and the optimum obtained from the strong learner approaches the actual design optimum. The strong learner used in this study is a committee machine [11], where multiple artificial neural networks are trained in parallel, and the final prediction is obtained by averaging the predictions of all the networks. While committee machines are used in this section, random forests and gradient-boosted regression trees would work well as strong learners too.

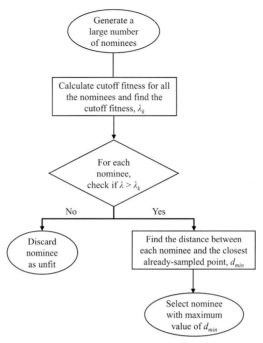

Figure 7.2 Flowchart showing the steps involved in obtaining new design points using the weak learner predictions.

At this point, the question arises: why do we need an ensemble model, made of separate strong and weak learners? The answer lies in the need for simultaneous exploration and exploitation. By querying the design space for exploitation using the strong learner, acceleration to the design optimum is achieved once the promising region has been located by the weak learner. Using a weak learner alone, on the contrary, would require a large number of simulations to reach the design optimum in the absence of any exploitation. On the other hand, using a strong learner alone will lead to premature convergence to local optima. However, by guiding the strong learner with the help of the weak learner, convergence to a local optimum is avoided.

2.3 Convergence criteria

An important element of design optimization is determining when to stop the iterative simulations. In other words, at what point are additional simulations unlikely to yield design improvements? Thus, defining criteria for convergence, which determine when to terminate the optimization process, is needed. For ActivO, changes observed in the weak learner response surface are used to determine convergence. Since the weak learner is meant to guide the exploration of the design surface, significant changes to the weak learner's predictions between successive iterations would suggest that ActivO is still exploring the design space and that the region that contains the design optimum has not been identified. Once the most promising regions of the design space have been identified, additional queries in the design space are unlikely to significantly change the weak learner's predictions. This behavior of the weak learner can be used to test for convergence. To do this, a set of monitor points are defined beforehand, and the weak learner predictions are recorded at these points as the optimization process progresses. ω, which is the maximum change in weak learner-predicted merits at these monitor points between the current iteration, iteration i, and the previous iteration, $i-1$, can be calculated. At iteration i, we can define the value of ω as:

$$\omega^i = \max \left| 100 \times \frac{\Phi^i - \Phi^{i-1}}{\Phi^{i-1}} \right| \% \qquad (7.1)$$

For termination of the optimization campaign, both the following conditions must be met:
a. Exploration is static: ω is less than some threshold value for five consecutive design iterations.
b. Exploitation is stagnant: The maximum design objective does not improve during these five iterations.

2.4 Dynamic exploration and exploitation

As mentioned in Section 2.1, the active learning process in ActivO combines query strategies from two different learners. In the basic algorithm (Fig. 7.3), the number of points obtained from the weak and strong learner query strategies are fixed throughout the optimization process. The strong learner provides one point, while the remaining points are obtained from the weak learner. Therefore, the balance between exploration and exploitation is fixed throughout the optimization process. On the contrary, it may be more beneficial to vary the number of points obtained from each learner as optimization progresses. This is because exploration involves gaining more information from regions where ground truth data are scarce, across the design space. Therefore, more exploration should be performed early on. On the other hand, the balance should tip in favor of exploitation as the optimization progresses and the promising region has been found.

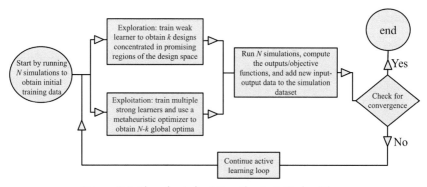

Figure 7.3 Flowchart depicting the ActivO algorithm.

At this stage, exploring the surface extensively may be wasteful of computing resources. Accordingly, a dynamic approach to exploration-exploitation balance has been adopted in ActivO. In this approach, the optimizer moves through three stages. The first stage occurs during the early iterations, where the weak and strong learners have relatively few training points. As such, there is high uncertainty in the design objective in most parts of the design space. At this stage, exploitation is very unlikely to yield any benefits. Therefore, all the query points are obtained using the weak learner (i.e., all points are used for exploration). This stage is referred to as the "extensive exploration" stage. In the second stage, known as "preliminary exploitation," there is a possibility (although uncertain) that the most promising region has been discovered, and thus preliminary exploitation begins, with 25% of the query points coming from the strong learner. In the final stage, it is expected that the promising region has been identified. At this stage, referred to as the "intensive exploitation" stage, equal numbers of query points are obtained from both learners. The transition between these three stages is controlled by ω, defined in the previous section. The three stages of dynamic exploration and exploitation are summarized in Table 7.1.

3. Case study 1: two-dimensional cosine mixture function

In this section, ActivO is used for optimizing a two-dimensional (2D) canonical problem, called the cosine mixture function [12], which is defined as:

$$z = 0.1(\cos 5\pi x + \cos 5\pi y) - (x^2 + y^2) \tag{7.2}$$

Table 7.1 Summary of the dynamic combination of weak learner and strong learner.

Optimization phase	p:N-p	Action
1. Extensive exploration	100:0	If ω increases, do nothing. If ω decreases, move to phase 2.
2. Preliminary exploitation	75:25	If ω increases, move to phase 1. If ω decreases, move to phase 3.
3. Intensive exploitation	50:50	If ω increases and new ω is above 5%, move to phase 2. If decreases or increases to a value less than 5%, do nothing.

A contour plot of the 2D cosine mixture function is shown in Fig. 7.4. In Eq. (7.2), $x \in [-1, 1]$, and $y \in [-1, 1]$ are the design variables, and maximizing z is the design objective. The cosine mixture function contains 25 local maxima and one global maximum (with a value of $z = 0.2$), located at $x = y = 0$.

In this study, optimization is carried out using a population size of five. In other words, five distinct points are queried in parallel at each design iteration. The initial design is obtained by drawing samples from a uniform distribution. As opposed to testing an optimizer on a problem once, it is helpful to run multiple trials, where the optimizer is reinitialized and the process is repeated to have representative performance projections. Therefore, 25 trials are performed for this problem.

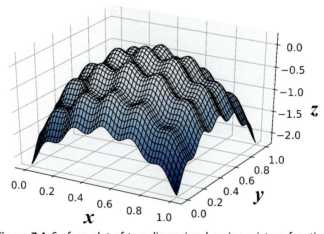

Figure 7.4 Surface plot of two-dimensional cosine mixture function.

For the 2D cosine mixture function, the performance of ActivO is compared against five other global optimizers, namely: microgenetic algorithm (μGA), particle swarm optimization (PSO), differential evolution (DE), a genetic algorithm using derivatives (GENOUD), and basin hopping (BH). μGA works by propagating desirable traits that lead to higher fitness values over successive design iterations while discarding individuals with undesirable traits. It was developed by Krishnakumar in 1989 [13] and is uniquely suited to problems that require small populations due to prohibitive computational costs. Even with small population sizes, μGA often succeeds in finding global optima, due to frequent restarts once the level of diversity in the population falls below a threshold. In this work, the population is reinitialized any time the variation in the binary traits fell below 5%. The second optimizer, PSO [14], is a metaheuristic technique that operates on the principle of swarm intelligence. Here, the position of each particle within the design space is updated sequentially, based on the best found by the entire swarm, the current particle's position, and the best position of the current particle. The inertia weight, which determines the balance between exploration and exploitation, is set to a constant value of 0.8. DE [15], a gradient-free global optimizer, combines existing design candidates to form new candidates while keeping the best candidate over successive generations. GENOUD [16] combines gradient-based quasi-newton methods with genetic algorithms to perform design optimization. Lastly, BH [17] involves random perturbation of coordinates, followed by local minimization, before rejecting or accepting the newly generated coordinates.

Fig. 7.5 shows the surface representation of the cosine mixture function as obtained from the strong and weak learners. As seen in the figure, the strong learner captures finer details including the local optima, while the weak learner ignores the multimodality of the design space. While ignoring the finer details, however, the weak learner correctly identifies that the points close to $x = y = 0$ have higher merit values. ActivO concentrates query points in regions of the design space where the weak learner (Fig. 7.5B) shows high fitness values. By probing the design space this way, the strong learner is able to find the exact location of the design optimum.

Fig. 7.6 shows the average maximum fitness value attained, as a function of the number of evaluations of the cosine mixture function. It can be seen that ActivO is more efficient compared to all the other optimizers. PSO, DE, and GENOUD fail to properly explore the surface. These optimizers

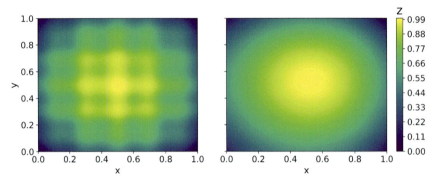

Figure 7.5 Strong learner (left) and weak learner (right) representations of the cosine mixture function.

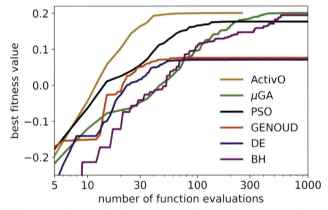

Figure 7.6 The evolution of the best fitness value for the cosine mixture problem using different global optimizers.

converge to local optima, leading to final objective values that are far from the global optimum. On average, PSO, DE, and GENOUD reach maximum fitness values of 0.17, 0.076, and 0.707, respectively. In contrast, ActivO and µGA both reach the theoretical maximum fitness value of 0.2, while BH reaches 0.194.

Another metric to consider is robustness, which concerns how reliable an optimizer is between different trials. This is because some optimizers may converge to the global optimum in some trials, but fail in other trials. A way to measure this is to examine the variance in the maximum fitness between different trials. A reliable optimizer is expected to have little variation between trials since it reaches the theoretical maximum consistently. Fig. 7.7 shows that in this regard, ActivO is superior to the other optimization

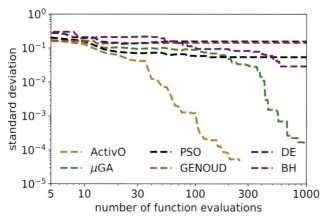

Figure 7.7 Standard deviation in the best fitness value for different optimization algorithms.

methods. After evaluating the cosine mixture function 100 times, the standard deviation of the maximum merit across the 25 trials is less than 1×10^{-3}. To attain such low variances, μGA takes 555 function evaluations. The other optimizers display much higher variances between different trials. PSO, DE, GENOUD, and BH show standard deviations of 0.059, 0.14, 0.16, and 0.03, even after 1000 function evaluations.

Another way to compare the optimizers is to examine the number of function evaluations needed to reach various fitness values, on average. This is shown in Fig. 7.8. Here, missing bars correspond to cases where an

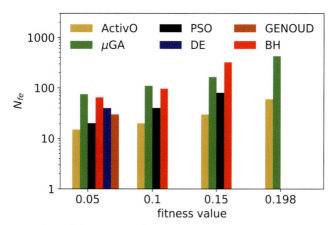

Figure 7.8 Number of function evaluations (N_{fe}) required to reach various fitness values.

optimizer fails to reach a given fitness value. Except for ActivO and μGA, all other optimizers fail to reach 0.198. In general, ActivO reaches all the thresholds 5–7x faster than μGA. PSO and BH are only able to reach fitness values of 0.15, on average. The worst-performing optimizers, DE and GENOUD, only reach fitness value of 0.05.

Fig. 7.9 shows the probability of locating the global optimum. To calculate this probability, the number of converged trials is divided by the total number of trials for a given number of function evaluations. Thus, the probability of convergence ranges from 0, when no trial has converged, to 1 when all the trials are converged. Using this metric, a cumulative histogram was constructed. Since the optimizers used are global optimizers, a tolerance of 0.002 is set, and optimizers are assumed to have reached the global optimum when they achieve a maximum fitness value of 0.198. It can be seen that PSO, DE, BH, and GENOUD, only reach a convergence probability of 80%, 4%, 96%, and 44%, even after 1000 function evaluations. In contrast, ActivO and μGA are able to reach a convergence probability of 100%, after 100 and 662 function calls, respectively. Therefore, for other algorithms besides ActivO and μGA, even after 1000 function evaluations, there is a chance that the optimum reached by the optimizer is not the global optimum. Failing to reach the global optimum after several function evaluations makes these optimizers unsuitable for

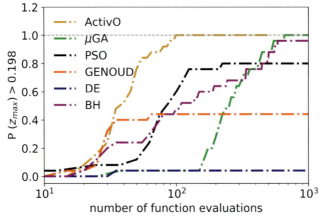

Figure 7.9 Improvement in the probability of convergence with increasing function evaluations.

CFD-driven optimization, where the function evaluations are compute-intensive. It should be noted that the poorly performing optimization algorithms are useful for many classes of problems. However, these optimizers appear to be unsuitable for CFD-driven optimization of IC engines due to the requirement of small population sizes over as few function evaluations as possible. In practice, each CFD simulation can require tens to hundreds of cores over several hours or days. Thus, running optimizers that involve large population sizes [$O(100)$ or higher] is infeasible. Therefore, an optimizer that is suited to engine design optimization needs a mechanism that enables it to sufficiently explore the design space even with small population sizes. As mentioned earlier, μGA does this by defining micro-convergence events. Here, the population (except for the fittest individual) is restarted once the population loses its diversity. ActivO, similarly, is able to avoid this problem, due to the presence of the weak learner. Since the weak learner does not capture the finer details within the design space, it can provide high-level details regarding promising regions, thus helping the strong learner queries escape local optima. In the next section, the two algorithms that demonstrated robustness over various trials for the cosine mixture function, ActivO and μGA, are compared for CFD-driven optimization of an IC engine.

4. Case study 2: computational fluid dynamics (CFD)-based engine optimization

In this section, CFD-driven optimization using ActivO is demonstrated for an IC engine application [18]. The engine operates on a low-octane gasoline-like fuel under medium-load conditions, with features such as exhaust gas recirculation and charged air cooling. A primary reference fuel blend containing 70% isooctane and 30% n-heptane was used as the fuel surrogate. Optimization was performed based on a single engine speed of 1375 rpm. The engine operating conditions are summarized in Table 7.2.

Here, design optimization is performed using a CFD model of the IC engine. The details of the computational model and its validation against experiments can be found in previous Refs. [5,18]. The list of design variables pertaining to fuel-air mixing, thermodynamic conditions, and fuel injection, are shown in Table 7.3.

Table 7.2 Summary of engine operating conditions [18].

Engine model	Cummins ISX15
Displacement	14.9 L
Bore	137 mm
Stroke	169 mm
Connecting rod	262 mm
Compression ratio	17.3:1
Engine speed	1375 rpm
Intake valve closing	−137°CA after top dead center (ATDC)
Exhaust valve opening	148°CA ATDC
Injection duration	15.58°CA
Mass of fuel injected	0.1245 g/cycle/cylinder
Fuel injection temperature	360 K
Global equivalence ratio	0.57

Table 7.3 List of design variables and their bounds.

Parameter	Description	Min	Max	Units
nNoz	Number of nozzle holes	8	10	—
TNA	Total nozzle area	1	1.3	—
Pinj	Injection pressure	1400	1800	bar
SOI	Start of injection timing	−11	−7	°CA ATDC
Nozzle angle	Nozzle half-inclusion angle	72.5	83.0	Degrees
EGR	EGR fraction	0.35	0.5	—
Tivc	IVC temperature	323	373	K
Pivc	IVC pressure	2.0	2.3	bar
SR	Swirl ratio	−2.4	−1	—

The design objective is to minimize the indicated specific fuel consumption (ISFC) while adhering to certain constraints. The design objective, called the merit function, is defined as:

$$\text{Merit} = 100 * \left\{ \frac{160}{\text{ISFC}} - 100 * f(\text{PMAX}) - 10 * f(\text{MPRR}) - f(M_{\text{soot}}) - f(M_{\text{NOx}}) \right\} \quad (7.3)$$

where

$$f(\text{PMAX}) = \begin{cases} \dfrac{\text{PMAX}}{220} - 1, & \text{PMAX} > 220 \text{ bar} \\ 0, & \text{PMAX} \leq 220 \text{ bar} \end{cases} \quad (7.4)$$

$$f(\text{MPRR}) = \begin{cases} \dfrac{\text{MPRR}}{15} - 1, & \text{MPRR} > 15 \text{ bar/CA} \\ 0, & \text{MPRR} \leq 15 \text{ bar/CA} \end{cases} \quad (7.5)$$

$$f(M_{\text{soot}}) = \begin{cases} \dfrac{M_{\text{soot}}}{0.0268} - 1, & M_{\text{soot}} > 0.0268 \text{ g/kWh} \\ 0, & M_{\text{soot}} \leq 0.0268 \text{ g/kWh} \end{cases} \quad (7.6)$$

$$f(M_{\text{NOx}}) = \begin{cases} \dfrac{M_{\text{NOx}}}{1.34} - 1, & M_{\text{NOx}} > 1.34 \text{ g/kWh} \\ 0, & M_{\text{NOx}} \leq 1.34 \text{ g/kWh} \end{cases} \quad (7.7)$$

In Eqs. (7.3)–(7.7), PMAX and MPRR are the maximum in-cylinder pressure and the maximum pressure rise rate, respectively. M_{soot} and M_{NOx} are the mass of soot and NO_x emissions, respectively, normalized by the power output of the engine. As long as the values of PMAX, MPRR, M_{soot}, and M_{NOx} remain below the respective constraints in Eqs. (7.4)–(7.7), these do not negatively impact the merit value (Eq. 7.3). Overall, ISFC bears an inverse relationship with the merit value. Achieving the design objective involves maximizing the merit, which corresponds to better fuel economy (lower ISFC). On the other hand, keeping PMAX and MPRR below their respective threshold is important to avoid the risk of mechanical failure.

In this section, the performance of ActivO is compared to that of μGA. For both ActivO and μGA, eight simulations, each corresponding to distinct design points, are run in parallel for each design iteration or generation. In this case, five different trials, initialized independently, are performed for ActivO. After completing the simulations, the relevant components of the merit in Eq. (7.3) (ISFC, PMAX, MPRR, M_{soot}, and M_{NOx}) are extracted from the simulation data. The values of these design metrics, along with their corresponding design variables are added to the training database before the ML models are updated. Next, the simulations for the next iteration are launched at query points obtained from the strong and weak learners. In this way, the data from CFD simulations are used to train the ML models, and the ML models, in turn, are used to guide data sampling for the next set of simulations. The proportion of query points obtained from both learners is varied according to the value of ω, as described in Section 2.4. Initially, all eight simulation points are determined by the weak learner, as preliminary exploration takes place. As the optimization process progresses, the number of strong learner points increases to two (preliminary exploitation), and then to four (intensive exploitation).

Fig. 7.10 depicts this evolution by showing how ω changes as a function of the number of CFD simulations completed. It can be seen that ω starts at a relatively high value, indicating that the promising region as identified by the weak learner is constantly changing. At this stage, newly collected information from CFD simulations is still altering the weak learner response surface significantly, and thus, the goal is to sufficiently explore the surface. As the optimization process progresses, the value of ω reduces due to the weak learner surface remaining relatively steady with respect to the number of iterations.

As done in the previous section, the best fitness achieved as a function of the number of simulations is shown in Fig. 7.11. In the figure, the dash-dotted lines represent individual trials. Due to the nondeterministic elements in the ActivO algorithm (initial data points, neural network training), all five trials converge differently. The points of convergence of each trial are depicted with the circle symbols in the plot. Table 7.4 shows how many iterations it takes to reach the threshold merit value of 104.0. In the best case among the five trials, ActivO reached 104.0 merit after running just 40 CFD simulations (corresponding to five design iterations). The worst-case took 136 CFD simulations (17 iterations), while the median case required 88 simulations (11 iterations) to reach 104.0.

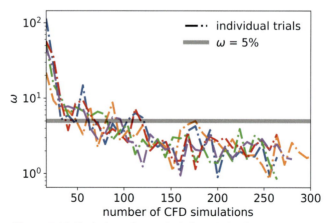

Figure 7.10 Evolution of ω as ActivO optimization progresses.

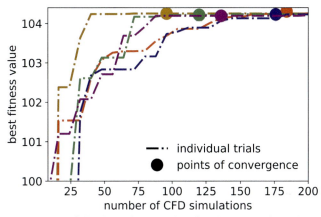

Figure 7.11 Evolution of the best fitness value for the internal combustion engine optimization problem obtained from five trials using ActivO.

Table 7.4 Summary of ActivO's performance on the internal combustion engine optimization problem.

	Number of design iterations to reach a fitness of 104	Number of design iterations to achieve convergence
Trial 1	5	12
Trial 2	16	23
Trial 3	9	15
Trial 4	17	22
Trial 5	11	17
Median	11	17
Mean	11.6	17.8

The convergence criteria were discussed in Section 2.3, where, in addition to the value of ω remaining sufficiently small, it is also required that the strong learner query points do not lead to an improvement in the maximum merit for five successive iterations. In this case, it is deemed useful to define a tolerance instead, where the optimization is terminated even if an improvement to the maximum merit occurred, as long as it is below some tolerance value. This is because, in some cases, very insignificant improvements can lead to the design optimizer running for several additional design iterations. In these cases, the cost of running tens or hundreds of more simulations outweighs the benefit of obtaining very small increments in the merit value. For this problem, the tolerance is set to

0.1, so that if ω is small, and the maximum merit does not increase by more than 0.1 for 5 iterations, the optimization loop is terminated. In this application, 0.1 is deemed an acceptable tolerance, but this may need to be adjusted based on the problem at hand and the level of accuracy of the CFD model used. The best, worst, and median cases required 96, 184, and 136 simulations, respectively, to achieve convergence. Table 7.5 shows the optimum parameters obtained from each ActivO trial. Some design parameters, such as nNoz, EGR, and Pivc, show little variation between the trials. Despite all trials reaching high merit values, the optimum found for other parameters, such as TNA, SOI, and SR, show considerable variation between various trials. This suggests that the merit value is very sensitive to the former list of parameters, while the latter have a broader range of acceptable values.

In understanding the operations of the weak and strong learners, it may be helpful to look into the merit values of the population members, as opposed to the best merit value alone. This is shown in Fig. 7.12. In this figure, the "+" symbols indicate that the point is a query from the strong learner, while the "o" symbols are from the weak learner queries of the design space. It can be seen that at the early stages, the query points (the "o" symbols) are exclusively from the weak learner. This changes progressively as the number of iterations increases. It can also be seen that in general, the weak learner query points seem to have a wider range of merit values, while the strong learner points seem to cluster around the best merits at a given design iteration. This further supports the overall hypothesis that the strong learner exploits the design space, while the weak learner explores it. It should also be noted that the merit values of the weak learner points seem

Table 7.5 Optimum design parameters.

Design parameter	Optimum parameters found				
	Trial 1	Trial 2	Trial 3	Trial 4	Trial 5
nNoz	10	10	10	10	10
TNA	1.07	1.09	1.12	1.18	1.09
Pinj	1481	1406	1415	1421	1413
SOI	−10.69	−9.66	−10.5	−10.3	−10.3
Nozzle Angle	157	155	156	157	157
EGR	0.45	0.44	0.45	0.45	0.45
Tivc	323	323	323	323	323
Pivc	2.3	2.3	2.3	2.3	2.3
SR	−1.84	−1.93	−1.98	−2.0	−2.0

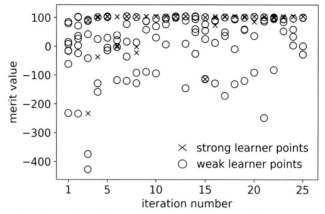

Figure 7.12 Population-based figure showing merit values of the points selected from the weak and strong learners.

to be higher toward the latter iterations compared to the earlier stages, suggesting that the weak learner learns to explore more promising regions as the optimization progresses.

Next, the performance of ActivO is compared to that of μGA as shown in Fig. 7.13. Here, the average performance of ActivO is compared to a single trial of μGA. A single trial of μGA is used for reasons that will soon be evident. On average, ActivO requires 88 CFD simulations to reach 104.0, while μGA takes 464 simulations. In contrast to the numbers presented in Table 7.2, μGA takes 784 simulations to reach convergence. Table 7.4

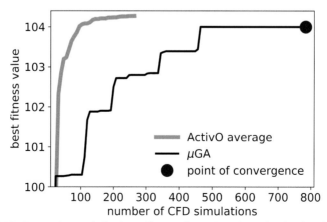

Figure 7.13 Comparison of the best fitness from ActivO and μGA for the engine optimization problem.

compares the performances of ActivO, ML-GA [5] (using a Super Learner surrogate coupled with a genetic algorithm), and μGA (coupling the μGA directly with CFD). It can be seen that both ML-GA and μGA utilize significantly more simulations compared to ActivO. Therefore, this demonstrates the capability of ActivO to enable computational savings for complex CFD-driven design optimization. If all the data from CFD can be generated in parallel, ML-GA would have the least runtime, but the amount of computing resources required to do this would be significant. Running hundreds of CFD simulations in parallel to train the Super Learner may be prohibitive for many users in industry and academia. In terms of core-hour usage, ActivO consumes 21,000 core-hours in total, while μGA and ML-GA consume 96,000 and 39,000 core-hours, respectively. With the μGA taking months and several core-hours for each optimization trial, performing multiple trials to test its performance is computationally infeasible. However, in a separate study [6], its performance was tested on a surrogate surface obtained by training a Super Learner model using over 2000 CFD simulations. Since the training data were obtained from the same engine configuration considered here, it can be assumed that the surrogate model thus obtained approximates the actual design space. In that study, μGA required 640 simulations, on average, to reach a merit 104.0, which gives an approximate indication of how it would perform if multiple trials were to be run. A summary of the performances of ML-GA, μGA, and ActivO is shown in Table 7.6.

Table 7.4 also shows the improvements to ISFC obtained using ActivO, μGA, and ML-GA. ISFC is shown because all three optimizers, at the design optimum, keep the constraints in Eqs. (7.4)–(7.7) below the required thresholds, causing them to have no adverse effects on the merit. ActivO reaches the lowest ISFC at 153.6 g/KWh, compared to the baseline 156.5 g/KWh, corresponding to savings of 2.9 g/KWh. ML-GA and μGA lead to smaller savings of 2.53 g/KWh and 2.65 g/KWh, respectively. Thus, it can be said that overall, ActivO leads to more efficient IC engine designs, while utilizing significantly lower computational resources.

Table 7.6 Comparison of the performances of ML-GA, μGA, and ActivO.

	Baseline	ML-GA	μGA	ActivO
Optimum ISFC (g/KWh)	156.53	153.97	153.85	153.6
Optimum merit	102.2	103.91	104.0	104.14
Number of simulations	–	250	640	136

It should be noted that although this engine optimization demonstration case only considers design parameters associated with fuel injection and initial flow/thermodynamic conditions, ActivO can be readily extended to optimization problems involving other parameters, such as piston bowl geometry and multiple speed-load conditions. In addition, ActivO can also be coupled with experiments for more efficient and faster engine calibration.

5. Conclusions

In this chapter, an active ML approach to sequential CFD-driven engine design optimization was described. ActivO is an adaptive surrogate-based approach that combines query strategies based on a strong learner and a weak learner to guide optimization toward the global design optimum. A mechanism for assessing convergence, as well as dynamically adjusting the balance between exploration and exploitation was introduced. As a preliminary test case, ActivO was tested on a 2D canonical test surface, called the cosine mixture function. The cosine mixture function contains two design variables, a single objective and global optimum, and 25 local optima that can potentially trap the optimizer. For this problem, ActivO was compared to five other optimizers, showing that on average, ActivO led to better fitness values while utilizing fewer function evaluations than others. Besides ActivO, the optimizers tested were a μGA, PSO, DE, GENOUD, and BH. Among all the optimizers tested, only ActivO and μGA successfully found the global optimum over all the trials. However, all of ActivO's trials found the global optimum within 100 function evaluations, as opposed to 662 for μGA.

Based on the results of the cosine mixture function, only ActivO and μGA were applied to the optimization of a compression-ignition IC engine operating on gasoline-like fuel. The goal was to minimize fuel consumption while adhering to constraints associated with emissions and engine mechanical limits by optimizing nine design variables relating to thermodynamic conditions, in-cylinder flow, and fuel injection. ActivO was compared to μGA, showing that on average, ActivO was about 5–7 times faster in achieving convergence. In general, this can provide immense benefits to the area of engine design optimization. More complex optimization endeavors, involving more design variables and multiple speed-load conditions can be carried out using ActivO. Overall, ActivO can significantly shorten the design cycle in the quest for cleaner and more efficient IC engines.

Acknowledgments

The submitted manuscript has been created by UChicago Argonne, LLC, Operator of Argonne National Laboratory (Argonne). The U.S. Government retains for itself, and others acting on its behalf, a paid-up nonexclusive, irrevocable world-wide license in said article to reproduce, prepare derivative works, distribute copies to the public, and perform publicly and display publicly, by or on behalf of the Government. This work was supported by the U.S. Department of Energy, Office of Science under contract DE-AC02-06CH11357. The research work was funded by the Department of Energy Office of Technology Transitions (OTT) and Vehicle Technologies Office (VTO). The authors acknowledge the computing resources available through the "Blues" and "Bebop" high-performance computing clusters operated by the Laboratory Computing Resource Center (LCRC) at Argonne National Laboratory utilized for this work.

References

[1] Badra J, Sim J, Pei Y, Viollet Y, Pal P, Futterer C, et al. Combustion system optimization of a light-duty GCI engine using CFD and machine learning. SAE Technical Paper. Society of Automotive Engineers; 2020, 2020-01-1313.
[2] Badra JA, Khaled F, Tang M, Pei Y, Kodavasal J, Pal P, et al. Engine combustion system optimization using CFD and machine learning: a methodological approach. J Energy Resour Technol 2020;143(2):022306. https://doi.org/10.1115/1.4047978. JERT-20-1594.
[3] Bertram AM, Kong S-C. Computational optimization of a diesel engine calibration using a novel SVM-PSO method. SAE Technical Paper. 2019.
[4] Kavuri C, Kokjohn SL. Exploring the potential of machine learning in reducing the computational time/expense and improving the reliability of engine optimization studies. Int J Engine Res 2018. 1468087418808949.
[5] Moiz A, Pal P, et al. A machine learning-genetic algorithm (ML-GA) approach for rapid optimization using high-performance computing. SAE Int J Commer Veh 2018;11(5):291−306. https://doi.org/10.4271/2018-01-0190.
[6] Owoyele O, Pal P. A novel active optimization approach for rapid and efficient design space exploration using ensemble machine learning. J Energy Resour Technol 2020;143(3):032307. https://doi.org/10.1115/1.4049178. JERT-20-1642.
[7] Owoyele O, Pal P. A novel machine learning-based optimization algorithm (ActivO) for accelerating simulation-driven engine design. Appl Energy 2021;285. https://doi.org/10.1016/j.apenergy.2021.116455.
[8] Broatch A, Novella R, Gomez-Soriano J, Pal P, Som S. Numerical methodology for optimization of compression-ignited engines considering combustion noise control. SAE Int J Engines 2018;11(6):625−42. https://doi.org/10.4271/2018-01-0193.
[9] Pei Y, Pal P, et al. CFD-guided combustion system optimization of a gasoline range fuel in a heavy-duty compression ignition engine using automatic piston geometry generation and a supercomputer. SAE Int J Adv Curr Pract Mobil 2019;1(1):166−79. https://doi.org/10.4271/2019-01-0001.
[10] Drucker H, Burges CJ, Kaufman L, Smola AJ, Vapnik V. Support vector regression machines. Adv Neural Inf Process Syst 1997:155−61.
[11] Haykin S. A comprehensive foundation. Neural Netw 2004;2:41.
[12] Breiman L, Cutler A. A deterministic algorithm for global optimization. Math Program 1993;58:179−99.

[13] Krishnakumar K. Micro-genetic algorithms for stationary and non-stationary function optimization. In: Intelligent Control and Adaptive Systems: International Society for Optics and Photonics; 1990. p. 289—97.
[14] Eberhart R, Kennedy J. Particle swarm optimization. In: Proceedings of the IEEE international conference on neural networks: Citeseer; 1995. p. 1942—8.
[15] Storn R, Price K. Differential evolution—a simple and efficient heuristic for global optimization over continuous spaces. J Global Optim 1997;11:341—59.
[16] Sekhon JS, Mebane Jr WR. Genetic optimization using derivatives. Polit Anal 1998:187—210.
[17] Wales DJ, Doye JP. Global optimization by basin-hopping and the lowest energy structures of Lennard-Jones clusters containing up to 110 atoms. J Phys Chem 1997;101:5111—6.
[18] Pal P, Probst D, Pei Y, et al. Numerical investigation of a gasoline-like fuel in a heavy-duty compression ignition engine using global sensitivity analysis. SAE Int J Fuels Lubr 2017;10(1):56—68. https://doi.org/10.4271/2017-01-0578.

SECTION 3

Artificial Intelligence to predict abnormal engine phenomena

CHAPTER 8

Artificial-intelligence-based prediction and control of combustion instabilities in spark-ignition engines

Bryan Maldonado[1], Anna Stefanopoulou[2] and Brian Kaul[1]

[1]Oak Ridge National Laboratory, Knoxville, TN, United States; [2]Department of Mechanical Engineering, University of Michigan, Ann Arbor, MI, United States

1. Introduction

1.1 Artificial intelligence applications to engine controls

Control strategies for internal combustion engines have become increasingly sophisticated in recent years, with a growing interest in predictive model-based strategies supplanting simpler table-based and proportional-integral-derivative (PID)-style reactive approaches [1]. Modern control methods rely on accurate physics-based models to predict the behavior of the system and optimize the control actions. Even with the recent advances in computational power, the complexity of many physical systems, such as the combustion process inside an engine, requires first principles modeling resulting in high-fidelity simulations assisted by high-performance computing facilities [2], rendering them impractical for real-time control implementation. With this increase in model complexity, artificial intelligence (AI) methods are becoming more attractive, potentially enabling model-free data-driven prediction and real-time control, even where physics-based models may not be practical.

Indeed, many examples of applications of AI to various problems in engine controls can be found. AI has been used to speed up models for control-oriented purposes by replacing some costly physics calculations [3–5] to predict various phenomena including idle combustion stability [6], knock [7,8], homogeneous charge compression ignition (HCCI) combustion phasing [9,10], and partially premixed combustion behavior [11]. It has also been used to create virtual sensors for estimates of cylinder pressure [12,13], indicated torque [14], NO_x emissions [15,16], and misfire detection [17,18] and as a replacement for volumetric efficiency or altitude

Artificial Intelligence and Data Driven Optimization of Internal Combustion Engines
ISBN 978-0-323-88457-0
https://doi.org/10.1016/B978-0-323-88457-0.00006-0

© 2022 Elsevier Inc.
All rights reserved.

compensation look-up tables [19,20]. Direct AI-based control strategies have also been implemented, both for individual actuators/subsystems such as electronic throttle bodies [21,22], exhaust gas recirculation (EGR) valves [23], and fuel injectors [24] and to replace PID or model-based control strategies for engine setpoints including engine speed [25−28], air/fuel ratio [29−33], and torque [34−37]. Applications to reducing noise and vibrations during hybrid vehicle engine restarts [38,39] and overall setpoint optimization [40,41] have also been demonstrated.

It is important to remember, though, that AI is not a panacea or a magic wand; rather, it is a family of automated methods that can be used for generating empirical models, identifying statistical patterns and correlations in data, and the like. The automation of these functions can be a powerful enabler for adaptive control strategies, but like all AI applications, it does have certain limitations. In particular, AI classifiers and regression models (which could be thought of as automated curve fits) can be quite good at interpolating within the range of training data that they were developed with, but are limited in extrapolating beyond the training region. Care should thus be taken to use AI-based empirical models within the range for which they are valid, but not to extrapolate beyond that. Overfitting is also a potential pitfall, just as it is when applying high-order polynomial regression without AI. Additionally, it is not generally possible to determine why an AI-based model or controller makes a given prediction or decision. This interpretability challenge does not necessarily impede the performance of AI in a given application, but limits the ability to use AI-based models for improving the understanding of the underlying physics in the way a physics-based model would. The nontransparent nature of AI-based control can also be problematic when analyzing the stability and robustness of the closed-loop system since classical methods, such as Lyapunov stability, are based on input-to-state functions rather than input-to-output relationships.

Many of these limitations can be somewhat mitigated by using AI components within a physics-informed framework. If pure AI approaches are thought of as "black-box" and first principles physics-based models or control strategies are thought of as "white-box," the proposed hybrid strategy could be considered a "gray-box" approach. Each of these approaches has its pros and cons, which will be highlighted in this chapter. White-box modeling approaches are generally realistic and accurate due to having a physical basis, while also being computationally expensive (often prohibitively so); if the underlying physics are poorly understood, it will not generally be possible to develop such a model. Black-box approaches are typically computationally inexpensive and easily implemented if sufficient training data exist to learn the system dynamics, but as mentioned previously, are prone to yield unrealistic predictions if extrapolating beyond the operating conditions of the initial training data and lack the interpretability and explainability of a physics-based approach. Gray-box approaches, consisting of the utilization of AI within a physics-informed structure, aim to synergistically combine physics-based and AI approaches in a complementary fashion.

Along with more sophisticated controls, several advanced combustion strategies, which are limited by abnormal combustion/combustion instability phenomena, are now employed to improve fuel efficiency. The combustion stability limits encountered pose constraints to further engine efficiency improvements. Such limits can be encountered at different conditions ranging from low-load operation, such as cyclic variability in dilute spark-ignition (SI) or HCCI combustion and during cold start, to high-load operation, such as knock and preignition for SI combustion. Some of these abnormal/unstable combustion events are thought to be stochastic, while others have been observed to exhibit deterministic patterns and comprise dynamical—and often chaotic—systems. Approaches to prediction and control will be different for the two cases: in stochastic systems, prediction will be probabilistic and stability control will focus on minimizing the potential of excursions into unstable regions while maximizing efficiency; in deterministic dynamical systems, next-cycle predictions based on prior events—and active controls utilizing these predictions—are theoretically possible (though challenging). In both cases, AI methods offer potential for improvement on classical approaches. This chapter will illustrate how AI techniques can be used to improve the prediction or control of abnormal combustion phenomena using several case studies.

1.2 Dilute combustion instability background

Improved prediction and mitigation of various abnormal combustion phenomena would address barriers to engine efficiency, and AI techniques are likely to yield benefits in multiple areas. Dilute SI combustion instabilities have been studied more thoroughly than other phenomena of interest, particularly as relates to the application of AI techniques, and will thus be used for the example case studies in this chapter. Accordingly, a brief overview of the phenomenon is given here.

Dilute combustion, accomplished either by excess air or with EGR, is a technologically proven and cost-effective way to improve fuel economy on a broad scale. The efficiency gain comes primarily at part load via improved thermodynamic (gamma) effects, decreased throttling losses, and decreased heat losses [42]. Stoichiometric operation with EGR, in particular, allows for improved efficiency while maintaining compatibility with three-way catalyst (TWC) aftertreatment systems to control emissions. However, at high dilution levels, combustion becomes unstable, and cycle-to-cycle variability (CCV) increases to unacceptable levels. Combustion stability may be limited either by flame initiation (misfires) [43,44] or propagation (partial burns) [45], depending on the spark timing [46].

Although previous research has focused on the estimation of the boundary of combustion stability [47], complete avoidance of misfires and partial burns remains an open problem. Misfire avoidance is of particular importance since such events cause drivability issues [48]. Additionally, a misfire occurrence rate of 2% or higher can cause TWC damage and increases emission levels over the threshold defined by on-board diagnostics regulations [49]. Therefore, a successful dilute combustion management strategy will maximize the levels of dilution while avoiding sporadic misfire and partial burns in order to guarantee maximum engine efficiency, improved emissions, and compliance with drivability standards.

Under conditions without severe partial burns and/or misfires, combustion events can be considered uncorrelated, and CCV is modeled mainly by a Gaussian random process [50,51]. At the dilute limit, however, it has been shown that combustion CCV exhibits deterministic patterns along with stochastic variations [52−56], mainly due to the nonlinear dependence of flame initiation and flame development on in-cylinder composition. The quantity and composition of residual gas on a given cycle directly impact the subsequent combustion event, serving as the mechanism for deterministic coupling between cycles [57]. This deterministic behavior could be minor, such as in stable SI operation, or significant, such as in dilute SI and HCCI operation. Fig. 8.1 shows the return maps for the normalized gross heat release Q_{gross} of three different operating conditions at increasing levels of determinism as EGR increases. The left plot represents a condition just before the dilute limit with a coefficient of variation (CoV) of the indicated mean effected pressure (IMEP) lower than the industry standard of 3%. Here, the return map shows a disperse pattern centered at the nominal value, typical of a random uncorrelated sequence. The center plot

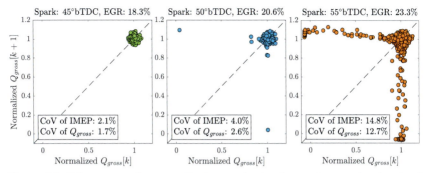

Figure 8.1 Return maps for normalized gross heat release at increasing levels of exhaust gas recirculation fraction and spark advance calibrated for optimal combustion phasing.

corresponds to a condition past the dilute limit where sporadic partial burns and misfires occur. Finally, the right plot shows a condition with very high CCV where the return map exhibits an asymmetric shape with respect to the $x = y$ diagonal, indicating the prior cycle correlation and the time irreversibility caused by nonlinear dynamics [58].

AI-based approaches have been attempted to improve modeling and control of dilute SI combustion stability. Three case studies have been selected to illustrate how AI techniques can enhance different components of a feedback control strategy. The case study in section 2 shows how to improve the modeling and prediction of combustion instabilities, required for model-based design of state estimators and controllers. The second case study in section 3 shows how to use neural network as control agents for next-cycle combustion stability control. Finally, the last case study in section 4 shows how AI can be used as an add-on strategy to optimize the operation of a combustion controller via online learning of the location of combustion stability limits.

2. Case study: artificial-intelligence-enhanced modeling of dilute spark-ignition cycle-to-cycle variability

This case study illustrates the use of AI for improving next-cycle prediction of dilute combustion instabilities. Although simple models, such as the one developed by Daw et al. [59], are preferred for control design and online implementation, they do not generate accurate predictions for specific engine cycles. More physically detailed computational fluid dynamics models, on the other hand, are more accurate at the cost of being far too computationally expensive for real-time control applications [60,61]. Therefore, AI-based methods are a promising approach to balance the trade-off between model complexity and prediction accuracy. In this case, a hybrid, gray-box model was developed which incorporates both physics-based and AI components in a control-oriented combustion model for next-cycle prediction of dilute SI CCV. In particular, the cycle-to-cycle combustion dynamics were informed by a physics-based approach, whereas the parameters of the model were identified from data using AI techniques. This novel AI-enhanced approach was compared against the earlier parametric model found in the literature. The simulation results illustrate the advantages of AI regarding modeling accuracy.

The physics-based (white-box) approach for the model was originally proposed by Daw et al. [59]. The dynamics are characterized by the in-cylinder composition, which corresponds to the residual gas from the previous combustion event and the fresh intake charge:

$$x_{k+1} = X_{\text{res}}[k] \begin{bmatrix} 1 - \eta_c[k] & 0 & 0 \\ -\text{AFR}_s \eta_c[k] & 1 & 0 \\ (1 + \text{AFR}_s)\eta_c[k] & 0 & 1 \end{bmatrix} x_k + \begin{bmatrix} 1 \\ 0 \\ 0 \end{bmatrix} u_k$$

$$+ \begin{bmatrix} 0 \\ 1 \\ \dfrac{X_{\text{EGR}}}{1 - X_{\text{EGR}}} \end{bmatrix} m_{\text{air,in}} \tag{8.1}$$

$$Q_{\text{gross}}[k] = \eta_c[k] M_{\text{fuel}}[k] Q_{\text{LHV}}. \tag{8.2}$$

The states $x_k = \begin{bmatrix} M_{\text{fuel}}[k] & M_{\text{air}}[k] & M_{\text{inert}}[k] \end{bmatrix}^T$ are the in-cylinder amounts of fuel, air, and inert gas. The model parameters are: (1) the stoichiometric air-to-fuel ratio AFR_s, (2) the lower heating value of the fuel Q_{LHV}, (3) the combustion efficiency η_c, and (4) the residual gas fraction X_{res}. The fresh charge is comprised of the injected fuel command $u_k = m_{\text{fuel,in}}[k]$, the fresh air $m_{\text{air,in}}$, and the EGR fraction X_{EGR}. Here, the cycle $k + 1$ starts at IVC, which is affected by the fuel injection u_k occurring before IVC. Fig. 8.2 illustrates the model.

The model parameters AFR_s and Q_{LHV} are constant properties of the fuel employed. The sharp drop of the combustion efficiency at the dilute

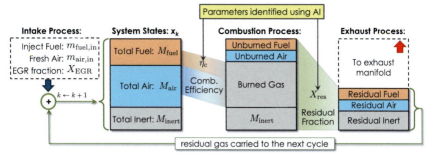

Figure 8.2 Physics-based control-oriented modeling of cycle-to-cycle combustion variability with artificial-intelligence-based components to increase model accuracy.

limit was modeled using a parametric sigmoid function with respect to the in-cylinder composition, quantified by the gas-fuel equivalence ratio [62]:

$$\lambda'[k] = \frac{M_{\text{air}}[k] + M_{\text{inert}}[k]}{M_{\text{fuel}}[k]} \cdot \frac{1}{\text{AFR}_s}. \tag{8.3}$$

Maldonado and Kaul [63,64] first extended the original model to account for the dependence of the residual gas fraction on the gross energy release of the preceding combustion event. Such a dependency was modeled by a parametric polynomial function. As indicated in Fig. 8.2, the sigmoid and the polynomial functions used to model $\eta_c[k]$ and $X_{\text{res}}[k]$, respectively, were replaced by AI methods to increase model accuracy.

In order to capture the average combustion behavior as well as the CCV observed, a nonparametric unsupervised kernel density estimator (KDE) was used to identify the conditional probability density function (PDF) of $\eta_c[k]$ and $X_{\text{res}}[k]$ using experimental data [65]. Using the KDE approach, the model parameters were treated as random variables with the following PDFs:

$$\eta_c[k] \sim \widehat{p}_{\eta|\lambda'}\left(\eta_c \mid \lambda'[k]\right) \quad \text{and} \quad X_{\text{res}}[k] \sim \widehat{p}_{X|Q}\left(X_{\text{res}} \mid Q_{\text{gross}}[k]\right). \tag{8.4}$$

The inverse cumulative distribution function (CDF) sampling was used to obtain the cycle-to-cycle values of $\eta_c[k]$ and $X_{\text{res}}[k]$ during simulations.

Fig. 8.3 shows the simulation results at an operating condition past the dilute limit where the sensitivity of the flame kernel development increases, causing sporadic partial burns and misfires. The top and middle rows show the estimated cycle-to-cycle values of combustion efficiency and residual

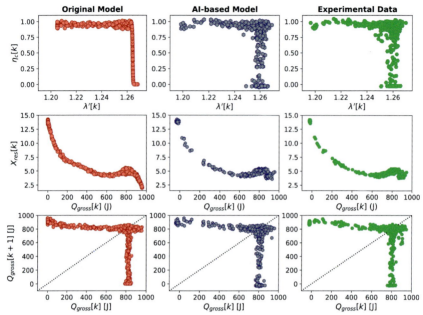

Figure 8.3 Comparison of cycle-to-cycle variability-related parameters between white-box model (left), artificial-intelligence-based model (center), and experimental data (right). *(Reproduced from Maldonado BP, Kaul BC. Control-oriented modeling of cycle-to-cycle combustion variability at the misfire limit in SI engines. In Proceedings of the ASME 2020 dynamic systems and control conference, page V002T26A001; 2020. doi: 10.1115/DSCC2020-3255 and Maldonado BP, Kaul BC, Schuman CD, Young SR, Mitchell JP. Next-cycle optimal fuel control for cycle-to-cycle variability reduction in EGR-diluted combustion. IEEE Control Syst Lett 2020;50(6):2204—2209. doi: 10.1109/LCSYS.2020.3046433.)*

gas fraction using the original white-box model (left column) and the AI-based gray-box model (center column). The right column shows the experimental data used to identify the model parameters. Both models show an appropriate amount of CCV for high combustion efficiency values. The original model, however, presents a sudden drop in η_c within a narrow interval, characteristic of the sigmoid function. The AI model, on the other hand, captures the variability observed in η_c and also the input-to-output uncertainty observed in the experimental data at lower combustion efficiency values. The overall characteristics of the residual gas fraction as a function of the gross heat release can be captured by either model. Nevertheless, the AI model is better at identifying discontinuities in the data where experimental values were not observed. The simple parametric nature of the original model makes it continuous, taking all possible values in the domain. The bottom row shows the return maps of gross heat release for the original model, AI-based model, and experimental data. The return maps were used to determine the model accuracy at capturing the next-cycle deterministic behavior observed, as well as the combustion CCV. As previously observed by Kaul et al. [53], at the dilute limit, combustion cycles with significantly low energy release leave a substantial amount of residual charge to be used during the next cycle, generating a subsequent higher-than-normal energy release. This effect is generally captured by both models, but the AI-based model more closely matches the experimental dynamics in the details. In order to quantify the performance of both models, the Cramér-von Mises distance between the experimental and simulated CDFs of Q_{gross} was calculated, indicating that the AI-based model is 48% more accurate than the original formulation.

3. Case study: neural networks for combustion stability control

This case study presents examples of using neural networks for next-cycle combustion stability control. The existence of deterministic effects in combustion instabilities beyond the dilute limit implies the possibility of mitigation through active control strategies, and a proof of concept has been demonstrated using simple proportional feedback control [66]. However, the application of modern control approaches is complicated both by the difficulty in predicting next-cycle combustion outcomes using physics-based models and the nonlinear sensitivity of combustion to charge composition that drives the chaotic behavior, which increases exponentially for dilution levels beyond stable combustion limits. Several examples have been demonstrated applying artificial neural networks (ANNs) to this application, with varying degrees of complexity. More recently, spiking neural networks (SNNs) have been applied to the dilute control problem which, distinct from traditional ANNs, incorporate the notion of time and causality.

3.1 Artificial neural networks

Neural network–based controllers were initially developed for lean combustion stability control by Vance et al. [67,68] and for EGR dilution by Singh et al. [69] and Vance et al. [70,71]. These controllers were designed with an ANN-based observer coupled with an ANN-based controller, as shown in Fig. 8.4. The observer network estimates the initial states of fuel and air mass, as well as the heat release based on the heat release of the prior cycle; the controller network then determines the desired fueling for the next cycle. These applications were designed and trained offline using the model of Daw et al. [59], then evaluated experimentally on a single-cylinder research engine. This approach was later extended by Shih et al. [72,73] to use an ANN-based adaptive-critic structure for online reinforcement learning, in place of the single-ANN controller.

These controllers were able to reduce the CoV of heat release when evaluated experimentally. Results for several lean equivalence ratios with the controller of Shih et al. [73] are shown in Table 8.1. Note that, for all

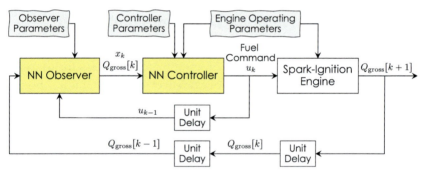

Figure 8.4 Neurocontroller structure using artificial neural networks for estimation and control.

Table 8.1 Coefficient of variation and fuel data for six operating conditions.

ϕ	CoV Uncontrolled	CoV Controlled	%CoV Change	%Fuel Change
0.89	0.0080	0.0077	−4.1	0.29
0.84	0.0090	0.0087	−3.3	0.11
0.79	0.0267	0.0221	−17.0	0.66
0.77	0.0475	0.0435	−8.3	0.48
0.75	0.1217	0.1071	−12.0	0.56
0.72	0.2373	0.2128	−10.3	0.48

Reproduced from Shih P, Kaul BC, Jagannathan S, Drallmeier JA. Reinforcement-learning-based output-feedback control of nonstrict nonlinear discrete-time systems with application to engine emission control. IEEE Trans Syst Man Cybern Part B (Cybernetics) 2009;390(5):1162−1179. doi: 10.1109/TSMCB.2009.2013272.

cases, there was a modest reduction in CoV as well as a small net increase in fueling. Due to the implementation of this control algorithm as a standalone controller with total control of the fuel injection quantity, it was not possible to entirely distinguish between the impact of cycle-resolved fueling changes and the impact of net enrichment on combustion stability.

3.2 Spiking neural networks

A more recent application of neural networks for dilute combustion stability control can be found in Maldonado et al. [74], where SNNs were implemented within a physics-based framework for real-time next-cycle dilute SI CCV control. SNNs take inspiration from biological systems and incorporate the notion of time into how they process information. Unlike traditional neural networks, synapses in an SNN can have different delay values, which determine how long it takes for the charge to travel along that synapse. Additionally, neurons accumulate charge over time and fire when a given threshold is reached. The firing of neurons in an SNN is asynchronous, depending on when each individual neuron reaches its threshold. The temporal processing capability of SNNs makes them better candidates at time series analysis and control tasks than traditional ANNs. The main advantage of SNNs regarding control tasks is their ability to accumulate information from all previous observations to inform the current action taken, rather than making a decision based on a fixed number of previous observations. Additionally, SNNs can be very efficiently implemented in neuromorphic hardware, where the computing hardware itself is made up of neurons and synapses [75]. In this study, however, the optimized SNNs were deployed on a μCaspian field-programmable gate array (FPGA) chip [76]. The low size, weight, and power demand of this hardware implementation are particularly attractive for powertrain applications.

The use of SNNs builds on the approach of earlier examples in a number of ways. First, the ANN observer is replaced with a physics-based estimation of the in-cylinder mass of fuel, air, and inert gas based on cylinder pressure data, as illustrated in Fig. 8.5. Additionally, a SNN is used for the controller instead of a traditional neural network. Rather than a manual network design as in prior work, these networks were developed and optimized using a genetic algorithm approach, using the gray-box model presented in the previous case study. The best-performing

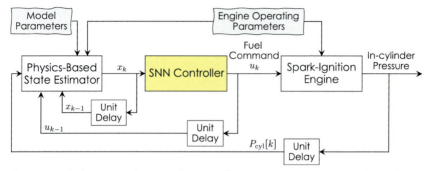

Figure 8.5 Spiking neural network controller structure using a gray-box design approach.

candidate networks were then experimentally evaluated together with an air-to-fuel ratio controller that allowed for closed-loop fuel trim to maintain stoichiometric operation.

The SNNs used for the dilute combustion control task were trained using an Evolutionary Optimization for Neuromorphic System (EONS) [77]. EONS utilizes a genetic or evolutionary approach to determine the optimized network topology (number of neurons and synapses, and how they are connected) and the network parameters (weights of synapses, delays of synapses, and thresholds for firing neurons). This process begins with an initial population of randomly initialized SNNs. Each SNN in the population is evaluated on how well it reduces combustion CCV by adjusting fuel quantity. Finally, the best-performing SNNs generate the next population based on reproduction operators such as duplication, recombination, and mutation. The EONS algorithm was repeated until a desired performance level was reached. A pictorial description of the EONS offline training methodology and subsequent real-time implementation is presented in Fig. 8.6. To speed up evolution, several EONS were run in parallel using the Summit supercomputer at Oak Ridge National Laboratory [78]. The optimized SNN was deployed in μCaspian and integrated with an engine control unit capable of next-cycle cylinder-pressure-based control [79]. Once the SNN-based controller was given full authority over the engine fuel injection, closed-loop experiments were conducted at EGR levels past the dilute limit, and spark advance was calibrated for optimal combustion phasing.

The left column of Fig. 8.7 summarizes the results at different levels of EGR and compares them against the open-loop (OL) baseline. In general, similar to the previous ANN example, the AI algorithm determined that

Artificial-intelligence-based prediction and control 197

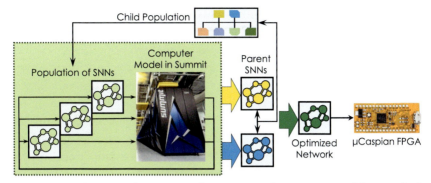

Figure 8.6 Evolutionary Optimization for Neuromorphic Systems workflow: Multiple spiking neural networks (SNNs) are run in parallel using the Summit supercomputer at ORNL. The best SNNs are used as parents to generate the next generation. Finally, the best SNN is deployed on the μCaspian field-programmable gate array.

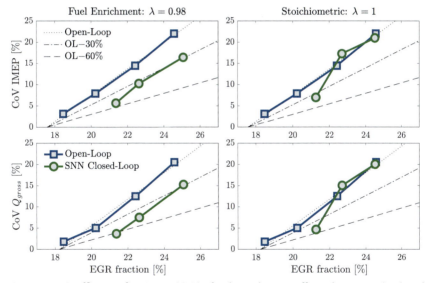

Figure 8.7 Coefficient of variation (CoV) of indicated mean effected pressure (top) and CoV of gross heat release (bottom) for conditions at increasing levels of exhaust gas recirculation operating under closed-loop spiking neural network (SNN)-based fuel control only (left) and a combination SNN-based and proportional-integral-based fuel control for stoichiometric combustion (right). *(Data from Maldonado BP, Kaul BC, Schuman CD, Young SR, Mitchell JP. Dilute combustion control using spiking neural networks. In SAE technical paper 2021-01-0534: SAE International; 2021. doi: 10.4271/2021-01-0534.)*

fuel enrichment can significantly reduce the combustion CCV. While this is a naïve solution to the control problem, the fact that the AI control agent learned the benefits of enriching the charge without being explicitly informed implies that they are capturing some of the underlying physical dynamics of the system. The property that small percentages of additional fuel can be enough to observe a significant reduction of CCV during high levels of EGR has been previously reported by Jatana and Kaul [80]. Therefore, the SNN controller increased the nominal fuel by approximately 2%, reducing the global air-fuel equivalence ratio from stoichiometric $\lambda = 1$ to rich $\lambda = 0.98$. Using fuel enrichment, the reduction of CoV near the dilute limit, around 21% EGR fraction, was as high as 60%. On the other hand, at the highest levels of EGR the relative CoV reduction was closer to 30%. In order to assess the effectiveness of AI at reducing CCV through next-cycle control while maintaining net stoichiometric conditions, a proportional integral (PI) controller was employed to maintain $\lambda = 1$ based on the exhaust oxygen sensor. This augmented strategy was conceived to separate the effects of next-cycle control from net fuel enrichment. The right column of Fig. 8.7 shows the summary of the experiments where the PI controller changed the nominal fuel quantity while the SNN controller performed small cycle-to-cycle adjustments. Under the new control strategy, a more modest reduction of CoV was observed for EGR < 23% while at the highest EGR conditions, the SNN controller was unable to reduce the CoV during stoichiometric conditions. This indicates that for very high EGR levels, the CoV reduction observed previously was primarily due to enrichment. Nonetheless, a 30% CoV reduction was observed close to the dilute limit at stoichiometric conditions, indicating that the AI algorithm successfully learned and utilized cycle-to-cycle combustion dynamics in addition to using enrichment.

This section exemplified the utilization of neural networks as state estimators and control agents. Both ANNs and SNNs were able to learn that a slight increase in the fuel quantity can have a significant impact on CCV reduction. Although valid, this solution is not compatible with current TWC operation, rendering it impractical. Nonetheless, the augmented control strategy pairing an SNN-based controller with a PI controller for $\lambda = 1$ operation showed that the AI strategy did learn the cycle-to-cycle deterministic dynamics of the system in order to reduce CCV through next-cycle control. These results show that SNNs are a promising approach for active combustion stability control. It is also apparent that care must be taken to avoid unintended naïve solutions such as enrichment.

4. Case study: learning reference governor for model-free dilute limit identification and avoidance

The final case study presents an AI method to dynamically learn the dilution limit after a boundary exploration period without the need for a physics-based model [81]. The AI algorithm was realized by a learning reference governor which is an add-on strategy applied to prestabilized systems in order to enforce input, output, and state constraints [82]. For automotive applications, load governors are popular strategies to limit the fueling and/or throttling rates during transients. For instance, a fuel governor has been previously implemented in an HCCI engine to avoid poor combustion [83]. Traditional reference governors, however, require a complete knowledge of the constraint boundary. Unfortunately, characterizing the dilute limit with a physics-based model requires an intense calibration effort and usually does not include the effects of engine aging.

Fig. 8.8 shows how the dilute limit changes based on a spark/EGR sweep at four different intake manifold pressures. The gray shaded area on each subplot corresponds to the region where the minimum indicated specific fuel consumption (ISFC) was achieved. Note that, across load conditions, the minimum ISFC was achieved right before the misfire limit. The variability of the boundary shape depicts the challenges of modeling the dilute limit at different loads. Fortunately, the empirical data showed the combustion phasing $CA50 = 7$ [CA deg] (solid blue line) is optimal across many conditions, and regulation to such a value using integral control would reduce the calibration problem [84]. However, this results in a constrained closed-loop combustion phasing control problem.

4.1 Constrained combustion phasing control problem

Fig. 8.9 shows the block diagram of the control strategy used to optimize combustion while avoiding the dilute limit during steady-state and transient conditions. The load demand originates from the driver and is executed by a drive-by-wire throttle valve. On the other hand, the spark advance and the EGR valve were manipulated by the controller. Feature identification is critical to determine the control targets. The combustion phasing CA50 is typically a robust indicator of optimal combustion [85] and has been widely used as a target for spark advance control using model-based control techniques [86–88] and AI-based methods [89,90]. Therefore, the spark advance and EGR valve were simultaneously adjusted by a linear quadratic

200 Artificial Intelligence and Data Driven Optimization of Internal Combustion Engines

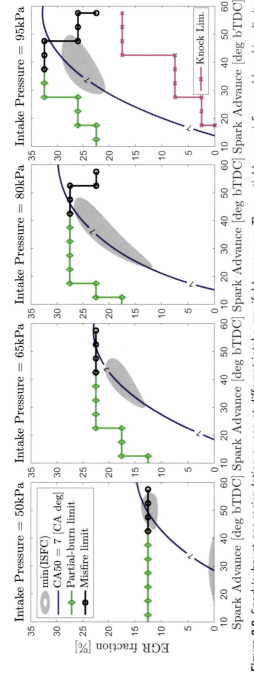

Figure 8.8 Spark/exhaust gas recirculation sweep at different intake manifold pressures. The partial burn, misfire, and knocking limits are highlighted. Minimum indicated specific fuel consumption is indicated by the gray shaded area and optimal combustion phasing by the blue solid line. (Reproduced from Maldonado BP, Li N, Kolmanovsky I, Stefanopoulou AG. Learning reference governor for cycle-to-cycle combustion control with misfire avoidance in spark-ignition engines at high exhaust gas recirculation–diluted conditions. Int J Eng Res 2020;210(10):1819–1834. doi: 10.1177/1468087420929109.)

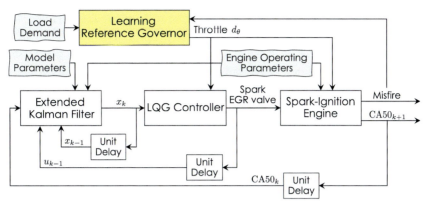

Figure 8.9 Learning reference governor add-on control strategy using an extended Kalman filter for state estimation and a linear quadratic Gaussian controller to adjust the spark advance and the exhaust gas recirculation valve.

Gaussian (LQG) controller to achieve the optimal CA50 and maximize EGR levels, as previously done by Maldonado et al. [91]. To avoid the dilute limit, however, the closed-loop system was augmented with a learning reference governor to filter the load demand, allowing smoother transients that can avoid constraint violation.

A physics-based model of the gas properties in the intake manifold, captured by a continuous-time nonlinear dynamic system, was used for control design [92]. Although the engine breathing dynamics are described by ordinary differential equations, the combustion events occur at discrete-time instants with a sample time equivalent to one engine cycle. Therefore, the engine model corresponds to a nonlinear, hybrid (continuous- and discrete-time), gray-box (data-driven components), stochastic dynamic system:

$$\frac{d}{dt}x = f(x, u, d_\theta) \quad (8.5)$$

$$CA50_k = g(x_k, u_k) + w_k. \quad (8.6)$$

Here, x corresponds to the system's states (intake manifold pressure and EGR fraction), u are the control actuators (spark advance and EGR valve), d_θ is the disturbance (throttle command), $CA50_k$ is the observation at cycle k, and w_k is the a Gaussian random variable corresponding to the stochastic component of CCV. Even though the dynamics $f(\cdot)$ were derived from gas properties, the observation $g(\cdot)$ was regressed from data. Such a regression can be done using traditional parametric techniques [93] or, more recently, using AI-based methods [94].

An adaptive extended Kalman filter was employed to reduce the propagation of CCV through the feedback loop. The adaptive nature allowed optimal filtering despite the lack of prior knowledge of the statistical properties of w_k. The statistical machine learning algorithm was realized using a finite impulse response filter during online estimation, as proposed by Maldonado et al. [95]. During closed-loop operation using the LQG controller, the dynamic response of the system presented undesirable behavior during throttle tip-outs where a sequence of misfire events was observed. The learning reference governor was able to identify the problematic dynamic behavior and adjust the throttle command to avoid it, without requiring any modeling of the dilute limit.

4.2 Learning reference governor for avoiding misfire events

Liu et al. [96] introduced the concept of a model-free learning reference governor applied to systems where constraint violations are undesirable but not catastrophic. In this application, the engine initially exhibited occasional misfires during throttle tip-outs and the reference governor learned, over time, how to avoid them. In order to achieve this goal, the throttle command was modified by the following feedback law:

$$\frac{d}{dt}v = \begin{cases} K_p \dfrac{d_\theta - v}{\max\{|d_\theta - v|, \eta\}} & \text{if } \varepsilon(x,v) < \Gamma(v) \\ 0 & \text{otherwise.} \end{cases} \quad (8.7)$$

Here, v is the modified throttle command, K_p and η are positive constants, and the functions $\varepsilon(\cdot)$, $\Gamma(\cdot)$ are such that the current state and constant throttle command pairs (x,v) satisfying $\varepsilon(x,v) \leq \Gamma(v)$ do not lead to constraint violations. The function $\varepsilon(\cdot)$ is typically chosen as $\|x - x^v\|_*$ where x^v is the steady-state value under the command v and $\|\cdot\|_*$ is an arbitrary vector norm. The function $\Gamma(\cdot)$ provides the machine learning component of the algorithm since it changes as the dilute limit is being identified, typically starting with a constant-value initial condition $\Gamma^0(v) = \overline{\gamma} \, \forall v$. During the learning period, the value of the function $\Gamma([v_{k-1}, v_k])$ is decreased every time a misfire is detected. Here, $[v_{k-1}, v_k]$ is to the interval corresponding time period preceding the misfire. The pair (ε, Γ) plays an analogous role to the dynamic safety margin defined by Nicotra and Garone, which determines how fast the load command v can evolve without running into constraint violation [97]. The machine learning technique is formally presented in Algorithm 1.

Algorithm 1 Learning reference governor (LRG).

1: **Input:** K_p, η, $\varepsilon(x,v)$, $\overline{\gamma} > 0$, $\lambda < 1$
2: **Output:** $\Gamma(v)$
3: **Initialization:** $n = 0$, $\Gamma^0(v) = \overline{\gamma} \forall v$
4: **while** Misfire **do**
5:     Perform a throttle tip-out.
6:     Use the LRG feedback policy in Eq. (8.7)
7:     **if** Cycle k misfires **then**
8:         Shrink $\Gamma(\cdot)$ at the cycle where misfire occurs:
$$\Gamma^{n+1}([v_{k-1}, v_k]) = \min\{\lambda \Gamma^n([v_{k-1}, v_k]), \varepsilon([(x_{k-1}, v_{k-1}), (x_k, v_k)])\}$$
9:         Enforce that $\Gamma^{n+1}(v)$ is continuous in v and satisfies:
$$0 < \Gamma^{n+1}(v) \le \Gamma^n(v) \quad \forall v$$
10:     **else** $\Gamma^{n+1}(\cdot) = \Gamma^n(\cdot)$
11:     **end if**
12:     $n \leftarrow n + 1$
13: **end while**

Fig. 8.10 shows the experimental response of the system to a throttle tip-out command before (left) and after (right) the learning phase was completed. Note that, before learning, the reference governor adjusted the throttle closing with an almost constant decrease rate. The EGR valve and the intake manifold pressure present similar responses. However, toward the end of the tip-out, two misfire events occurred, generating an abrupt response on spark advance and EGR valve. After this initial step, the function $\Gamma(\cdot)$ was updated according to the learning Algorithm 1. This procedure was repeated sequentially until, after a few iterations, the throttle tip-out command did not cause misfire events. After the learning was completed, note that the filtered throttle command v pauses momentarily toward the end of the tip-out, avoiding misfires. This happened when the condition $\varepsilon(x, v) \ge \Gamma(v)$ was met. The manipulation of the load demand by the learning reference governor slowed down the tip-out response, taking almost double the amount of time to reduce the engine load. Although undesirable for a conventional gasoline vehicle, this control strategy could be integrated in a hybrid electric vehicle where the torque generated during the longer tip-out can be used to charge the battery. The synergy between electrification and combustion control has the potential to enable a highly optimized engine operation for a variety of driving scenarios.

This case study showed that, in addition to use AI techniques to improve models and automatic controllers, AI techniques can also be used

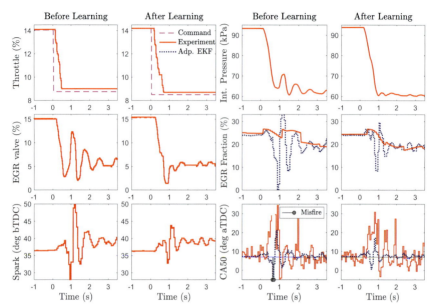

Figure 8.10 Closed-loop response of a throttle tip-out with the learning reference governor before learning (left) and after learning (right). *(Reproduced from Maldonado BP, Li N, Kolmanovsky I, Stefanopoulou AG. Learning reference governor for cycle-to-cycle combustion control with misfire avoidance in spark-ignition engines at high exhaust gas recirculation—diluted conditions. Int J Eng Res 2020;210(10):1819–1834. doi: 10.1177/1468087420929109.)*

as a supervisory strategy to regulate the driver demands when a constraint violation is predicted. By allowing the learning reference governor to filter the driver's load demand, aggressive transients which drive the system past the dilute limit can be avoided. The experiments confirmed the potential of AI-enhanced learning for identifying the dilute limit, maintaining operation within a feasible combustion region, and operating at high-efficiency combustion modes.

5. Summary

The three case studies showed in this chapter demonstrated the benefits of AI-learned patterns to fill-in, improve, or simplify physics-based models for reducing combustion CCV and identifying stability limits in SI dilute combustion. The synergy between data-driven AI and the physics-based understanding of the system allowed the development of next-cycle real-time control strategies, without requiring computationally expensive and complex combustion models. Moreover, each case study exemplifies how different components of a feedback control loop can be enhanced using AI, such as state estimators, control agents, and command governors; hence, illustrating that AI can be a powerful tool for enabling data-driven modeling and control.

As with other tools, judicious application is key to achieving good results and avoiding pitfalls such as extrapolating, overfitting data, or naïve solutions to control problems. Implementing AI-based models or control strategies in a physics-informed manner can mitigate these issues while taking advantage of AI's potential for enabling computationally inexpensive predictions and data-driven predictive control strategies.

AI approaches are also likely to be well-suited to a number of challenges associated with prediction and control of other abnormal combustion instabilities, beyond dilute SI combustion. Combustion stability during catalyst-heating operation at cold start in SI engines is a potential application that could likely make use of similar approaches to those used for combustion stability at the dilute combustion limit. Cyclic variations in HCCI and other related kinetically driven combustion modes have also been shown to exhibit nonlinear dynamics and chaos-like effects [98,99], and are thus also likely to benefit from a similar approach.

Other types of abnormal combustion events, such as SI knock or pre-ignition, might benefit from AI techniques as well. Further study is needed to determine whether identifiable precursors exist in the data to which AI prediction and control methods, like those described in this chapter, could be applied. In particular, unsupervised machine learning could potentially be used to identify correlations needed for predictive control.

References

[1] Re LD, Allgöwer F, Glielmo L, Guardiola C, Kolmanovsky I. Automotive model predictive control: models, methods and applications. Lecture notes in control and information sciences. London: Springer; 2010, ISBN 9781849960717.
[2] Jupudi RS, Finney CEA, Primus R, Wijeyakulasuriya S, Klingbeil AE, Tamma B, Stoyanov MK. Application of high performance computing for simulating cycle-to-cycle variation in dual-fuel combustion engines. In: SAE technical paper 2016-01-0798. SAE International; 2016. https://doi.org/10.4271/2016-01-0798.
[3] Papadimitriou I, Warner M, Silvestri J, Lennblad J, Tabar S. Neural network based fast-running engine models for control-oriented applications. In: SAE technical paper 2005-01-0072. SAE International; 2005. https://doi.org/10.4271/2005-01-0072.
[4] Egan D, Koli R, Zhu Q, Prucka R. Use of machine learning for real-time non-linear model predictive engine control. In: SAE technical paper 2019-01-1289. SAE International; 2019. https://doi.org/10.4271/2019-01-1289.

[5] Chen AS, Vorraro G, Turner M, Islam R, Herrmann G, Burgess S, Brace C, Turner J, Bailey N. Control-oriented modelling of a Wankel rotary engine: a synthesis approach of state space and neural networks. In: SAE technical paper 2020-01-0253. SAE International; 2020. https://doi.org/10.4271/2020-01-0253.

[6] Li X, Zouani A. Machine learning algorithm for the prediction of idle combustion uniformity. SAE Int J Adv Curr Pract Mobil 2019;1(4):1803−7. https://doi.org/10.4271/2019-01-1551.

[7] Petrucci L, Ricci F, Mariani F, Cruccolini V, Violi M. Engine knock evaluation using a machine learning approach. In: SAE technical paper 2020-24-0005. SAE International; 2020. https://doi.org/10.4271/2020-24-0005.

[8] Shen X, Zhang Y, Sata K, Shen T. Gaussian mixture model clustering-based knock threshold learning in automotive engines. IEEE ASME Trans Mechatron 2020;25(6):2981−91. https://doi.org/10.1109/TMECH.2020.3000732.

[9] Vaughan A, Bohac SV. Real-time, adaptive machine learning for non-stationary, near chaotic gasoline engine combustion time series. Neural Netw 2015;70:18−26. https://doi.org/10.1016/j.neunet.2015.04.007.

[10] Janakiraman VM, Nguyen X, Sterniak J, Assanis D. Identification of the dynamic operating envelope of HCCI engines using class imbalance learning. IEEE Trans Neural Netw Learn Syst 2015;26(1):98−112. https://doi.org/10.1109/TNNLS.2014.2311466.

[11] Li X, Yin L, Tunestal P, Johansson R. Learning based model predictive control of combustion timing in multi-cylinder partially premixed combustion engine. In: SAE technical paper 2019-24-0016. SAE International; 2019. https://doi.org/10.4271/2019-24-0016.

[12] Bizon K, Continillo G, Mancaruso E, Vaglieco BM. Towards on-line prediction of the in-cylinder pressure in diesel engines from engine vibration using artificial neural networks. In: SAE technical paper 2013-24-0137. SAE International; 2013. https://doi.org/10.4271/2013-24-0137.

[13] Bennett C, Dunne JF, Trimby S, Richardson D. Engine cylinder pressure reconstruction using crank kinematics and recurrently-trained neural networks. Mech Syst Signal Process 2017;85:126−45. https://doi.org/10.1016/j.ymssp.2016.07.015.

[14] Gani E, Manzie C. Indicated torque reconstruction from instantaneous engine speed in a six-cylinder SI engine using support vector machines. In: SAE technical paper 2005-01-0030. SAE International; 2005. https://doi.org/10.4271/2005-01-0030.

[15] Steyskal M, Olsen D, Willson B. Development of PEMS models for predicting NOx emissions from large bore natural gas engines. In: SAE technical paper 2001-01-1914. SAE International; 2001. https://doi.org/10.4271/2001-01-1914.

[16] De Cesare M, Covassin F. Neural network based models for virtual NO_x sensing of compression ignition engines. In: SAE technical paper 2011-24-0157. SAE International; 2011. https://doi.org/10.4271/2011-24-0157.

[17] Wu ZJ, Lee A. Misfire detection using a dynamic neural network with output feedback. In: SAE technical paper 980515. SAE International; 1998. https://doi.org/10.4271/980515.

[18] Chen SK, Mandal A, Chien L, Ortiz-Soto E. Machine learning for misfire detection in a dynamic skip fire engine. SAE Int J Eng 2018;110(6):965−76. https://doi.org/10.4271/2018-01-1158.

[19] Wu B, Filipi Z, Kramer DM, Ohi GL, Prucka MJ, DiValetin E. Using neural networks to compensate altitude effects on the air flow rate in variable valve timing engines. In: SAE technical paper 2005-01-0066. SAE International; 2005. https://doi.org/10.4271/2005-01-0066.

[20] Malaczynski GW, Mueller M, Pfeiffer J, Cabush D, Hoyer K. Replacing volumetric efficiency calibration look-up tables with artificial neural network-based algorithm for variable valve actuation. In: SAE technical paper 2010-01-0158. SAE International; 2010. https://doi.org/10.4271/2010-01-0158.
[21] Yang H, Liu LY, Wasacz B. Neural network based feedforward control for electronic throttles. In: SAE technical paper 2002-01-1149. SAE International; 2002. https://doi.org/10.4271/2002-01-1149.
[22] Al-Assadi S. Neural network-based model reference adaptive control for electronic throttle systems. In: SAE technical paper 2007-01-1628. SAE International; 2007. https://doi.org/10.4271/2007-01-1628.
[23] Krakowian KM, Skretowicz M. Application of artificial neural networks in exhaust gas recirculation systems. In: SAE technical paper 2020-01-2172. SAE International; 2020. https://doi.org/10.4271/2020-01-2172.
[24] Lucido M, Shibata J. Learning gasoline direct injector dynamics using artificial neural networks. In: SAE technical paper 2018-01-0863. SAE International; 2018. https://doi.org/10.4271/2018-01-0863.
[25] Salam FM, Gharbi AB. Temporal neuro-control of idle engine speed. In: Proceedings of the 1996 IEEE international symposium on intelligent control; 1996. p. 396–401. https://doi.org/10.1109/ISIC.1996.556234.
[26] Xue J, Gao Q, Ju W. Reinforcement learning for engine idle speed control. In: 2010 international conference on measuring technology and mechatronics automation, vol. 2; 2010. p. 1008–11. https://doi.org/10.1109/ICMTMA.2010.249.
[27] Libiao W, Jian F. Study of self-adaptive RBF neural network control method for the engine idle speed control. In: 2011 international conference on consumer electronics, communications and networks (CECNet). IEEE; 2011. p. 2633–6. https://doi.org/10.1109/CECNET.2011.5768667.
[28] Song E, Liu J, Ding S, Wang Y, Yao C. Study of neural network control algorithm in the diesel engine. In: SAE technical paper 2016-01-8086. SAE International; 2016. https://doi.org/10.4271/2016-01-8086.
[29] Asik JR, Peters JM, Meyer GM, Tang DX. Transient A/F estimation and control using a neural network. In: SAE technical paper 970619. SAE International; 1997. https://doi.org/10.4271/970619.
[30] Lenz U, Schroeder D. Air/fuel ratio control for direct injecting combustion engines using neural networks. In: SAE technical paper 1997-25-0094. SAE International; 1997. https://doi.org/10.4271/1997-25-0094.
[31] Arsie I, Pianese C, Sorrentino M. Development and real-time implementation of recurrent neural networks for AFR prediction and control. SAE Int J Passeng Cars Electron Electric Syst 2009;1(1):403–12. https://doi.org/10.4271/2008-01-0993.
[32] Zhai Y, Yu D. Neural network model-based automotive engine air/fuel ratio control and robustness evaluation. Eng Appl Artif Intell 2009;22(2):171–80. https://doi.org/10.1016/j.engappai.2008.08.001.
[33] Wong PK, Wong HC, Vong CM, Xie Z, Huang S. Model predictive engine air-ratio control using online sequential extreme learning machine. Neural Comput Appl 2016;27:79–92. https://doi.org/10.1007/s00521-014-1555-7.
[34] Liu D, Javaherian H, Kovalenko O, Huang T. Adaptive critic learning techniques for engine torque and air–fuel ratio control. IEEE Trans Syst Man Cybern Part B (Cybernetics) 2008;38(4):988–93. https://doi.org/10.1109/TSMCB.2008.922019.
[35] Müller R, Schneider B. Approximation and control of the engine torque using neural networks. In: SAE technical paper 2000-01-0929. SAE International; 2000. https://doi.org/10.4271/2000-01-0929.

[36] Gerasimov DN, Pshenichnikova EI. Neural network data-driven engine torque and air-fuel ratio control. In: 2012 16th IEEE mediterranean electrotechnical conference. IEEE; 2012. p. 524–7. https://doi.org/10.1109/MELCON.2012.6196487.
[37] Zeng W, Khalid MAS, Han X, Tjong J. A study on extreme learning machine for gasoline engine torque prediction. IEEE Access 2020;8:104762–74. https://doi.org/10.1109/ACCESS.2020.3000152.
[38] Radmilovic Z, Zehetner J, Watzenig D. Vibration comfort control for HEV based on machine learning. In: SAE technical paper 2014-01-2091. SAE International; 2014. https://doi.org/10.4271/2014-01-2091.
[39] Shimode K, Ishizaki K, Komada M. Machine learning based technology for reducing engine starting vibration of hybrid vehicles. In: SAE technical paper 2019-01-1450. SAE International; 2019. https://doi.org/10.4271/2019-01-1450.
[40] Khatri DS, Shiv Kumar B. Feasibility study of neural network approach in engine management system in S.I. Engine. In: SAE technical paper 2000-01-1426. SAE International; 2000. https://doi.org/10.4271/2000-01-1426.
[41] Shuvom MAA, Haq MZ. Development and analysis of adaptive neural network control for a cybernetic intelligent 'iGDI' engine. In: SAE technical paper 2015-01-0157. SAE International; 2015. https://doi.org/10.4271/2015-01-0157.
[42] Alger T, Gingrich J, Roberts C, Mangold B. Cooled exhaust-gas recirculation for fuel economy and emissions improvement in gasoline engines. Int J Eng Res 2011;12(3):252–64. https://doi.org/10.1177/1468087411402442.
[43] Aleiferis PG, Taylor AMKP, Whitelaw JH, Ishii K, Urata Y. Cyclic variations of initial flame kernel growth in a Honda VTEC-E lean-burn spark-ignition engine. In: SAE technical paper 2000-01-1207. SAE International; 2000. https://doi.org/10.4271/2000-01-1207.
[44] Lian H, Martz JB, Maldonado BP, Stefanopoulou AG, Zaseck K, Wilkie J, et al. Prediction of flame burning velocity at early flame development time with high exhaust gas recirculation and spark advance. J Eng Gas Turbines Power 2017;139(8). https://doi.org/10.1115/1.4035849. 0 082801–082801–9.
[45] Cha J, Kwon J, Cho Y, Park S. The effect of exhaust gas recirculation (EGR) on combustion stability, engine performance and exhaust emissions in a gasoline engine. KSME Int J 2001;15(10):1442–50. https://doi.org/10.1007/BF03185686.
[46] Quader AA. What limits lean operation in spark ignition engines-flame initiation or propagation?. In: SAE technical paper 760760. SAE International; 1976. https://doi.org/10.4271/760760.
[47] Maldonado BP, Stefanopoulou AG. Cycle-to-Cycle feedback for combustion control of spark advance at the misfire limit. J Eng Gas Turbines Power 2018;140(10):102812. https://doi.org/10.1115/1.4039728.
[48] Quader AA. Lean combustion and the misfire limit in spark ignition engines. In: SAE technical paper 741055. SAE International; 1974. p. 3274–96. https://doi.org/10.4271/741055.
[49] Gilles T. Automotive service: inspection, maintenance, repair. Cengage Learning; 2015, ISBN 9781305445932.
[50] Maldonado BP, Freudenberg JS, Stefanopoulou AG. Stochastic feedback combustion control at high dilution limit. In: 2018 annual American control conference (ACC); 2018. p. 1598–603. https://doi.org/10.23919/ACC.2018.8431020.
[51] Maldonado BP, Stefanopoulou AG. Linear stochastic modeling and control of diluted combustion for SI engines. IFAC-PapersOnLine 2018;51(31):99–104. https://doi.org/10.1016/j.ifacol.2018.10.019.

[52] Finney CEA, Kaul BC, Daw CS, Wagner RM, Edwards KD, Green Jr BJ. Invited review: a review of deterministic effects in cyclic variability of internal combustion engines. Int J Eng Res 2015;16(3):366–78. https://doi.org/10.1177/1468087415572033.
[53] Kaul B, Wagner R, Green J. Analysis of cyclic variability of heat release for high-EGR GDI engine operation with observations on implications for effective control. SAE Int J Eng 2013;6(1):132–41. https://doi.org/10.4271/2013-01-0270.
[54] Maldonado BP, Stefanopoulou AG. Non-equiprobable statistical analysis of misfires and partial burns for cycle-to-cycle control of combustion variability. In: Proceedings of the ASME 2018 internal combustion engine division fall technical conference, page V002T05A003. ASME; 2018. https://doi.org/10.1115/ICEF2018-9540.
[55] Maldonado B, Stefanopoulou A, Scarcelli R, Som S. Characteristics of cycle-to-cycle combustion variability at partial-burn limited and misfire limited spark timing under highly diluted conditions. In: Proceedings of the ASME 2019 internal combustion engine division fall technical conference, page V001T03A018. ASME; 2019. https://doi.org/10.1115/ICEF2019-7256.
[56] Stiffler R, Kaul B, Drallmeier J. Cyclic dynamics of misfires and partial burns in a dilute spark-ignition engine. Proc Inst Mech Eng - Part D J Automob Eng 2021;235(2–3):333–45. https://doi.org/10.1177/0954407020964004.
[57] Daw CS, Kennel MB, Finney CEA, Connolly FT. Observing and modeling nonlinear dynamics in an internal combustion engine. Phys Rev E 1998;57:2811–9. https://doi.org/10.1103/PhysRevE.57.2811.
[58] Green JB, Daw CS, Armfield JS, Finney CEA, Wagner RM, Drallmeier JA, Kennel MB, Durbetaki P. Time irreversibility and comparison of cyclic-variability models. SAE Trans 1999;108:355–62. https://doi.org/10.4271/1999-01-0221.
[59] Daw CS, Finney CEA, Green JB, Kennel MB, Thomas JF, Connolly FT. A simple model for cyclic variations in a spark-ignition engine. In: SAE technical paper 962086. SAE International; 1996. https://doi.org/10.4271/962086.
[60] Finney CEA, Stoyanov MK, Pannala S, Daw CS, Wagner RM, Edwards KD, Webster CG, Green Jr JB. Application of high performance computing for simulating the unstable dynamics of dilute spark-ignited combustion. In: International conference on theory and application of nonlinear dynamics (ICAND 2012). Springer; 2012. p. 259–70. https://doi.org/10.1007/978-3-319-02925-2_23.
[61] Scarcelli R, Sevik J, Wallner T, Richards K, Pomraning E, Senecal PK. Capturing cyclic variability in exhaust gas recirculation dilute spark-ignition combustion using multicycle RANS. J Eng Gas Turbines Power 2016;138(11). https://doi.org/10.1115/1.4033184.
[62] Shayler PJ, Winborn LD, Hill MJ, Eade D. The influence of gas/fuel ratio on combustion stability and misfire limits of spark ignition engines. In: SAE technical paper 2000-01-1208. SAE International; 2000. https://doi.org/10.4271/2000-01-1208.
[63] Maldonado BP, Kaul BC. Control-oriented modeling of cycle-to-cycle combustion variability at the misfire limit in SI engines. In: Proceedings of the ASME 2020 dynamic systems and control conference, page V002T26A001; 2020. https://doi.org/10.1115/DSCC2020-3255.
[64] Maldonado BP, Kaul BC. Evaluation of residual gas fraction estimation methods for cycle-to-cycle combustion variability analysis and modeling. Int J Eng Res 2021:1–16. https://doi.org/10.1177/1468087420983087.
[65] Maldonado BP, Kaul BC, Schuman CD, Young SR, Mitchell JP. Next-cycle optimal fuel control for cycle-to-cycle variability reduction in EGR-diluted combustion. IEEE Control Syst Lett 2020;5(6):2204–9. https://doi.org/10.1109/LCSYS.2020.3046433.

[66] Daw CS, Green JB, Wagner RM, Finney CEA, Davis LI, Feldkamp LA, Hoard JW, Yuan F, Connolly FT. Controlling cyclic combustion variations in lean-fueled spark-ignition engines. In: AIP Conference Proceedings. 6220; 2002. p. 265–77. https://doi.org/10.1063/1.1487542. 1.

[67] Vance JB, He P, Kaul B, Jagannathan S, Drallmeier JA. Neural network-based output feedback controller for lean operation of spark ignition engines. In: 2006 American control conference; 2006. p. 8. https://doi.org/10.1109/ACC.2006.1656497.

[68] Vance JB, Kaul BC, Jagannathan S, Drallmeier JA. Output feedback controller for operation of spark ignition engines at lean conditions using neural networks. IEEE Trans Contr Syst Technol 2008;16(2):214–28. https://doi.org/10.1109/TCST.2007.903368.

[69] Singh A, Vance JB, Kaul B, Drallmeier J, Jagannathan S. Neural network control of spark ignition engines with high EGR levels. In: The 2006 IEEE international joint conference on neural network proceedings; 2006. p. 4978–85. https://doi.org/10.1109/IJCNN.2006.247201.

[70] Vance JB, Singh A, Kaul BC, Jagannathan S, Drallmeier JA. Neural network controller development and implementation for spark ignition engines with high EGR levels. IEEE Trans Neural Network 2007;18(4):1083–100. https://doi.org/10.1109/TNN.2007.899199.

[71] Vance J, Kaul B, Jagannathan S, Drallmeier J. Neuro emission controller for minimising cyclic dispersion in spark ignition engines with EGR levels. Int J Gen Syst 2009;38(1):45–72. https://doi.org/10.1080/03081070802193028.

[72] Shih P, Kaul BC, Jagannathan S, Drallmeier JA. Reinforcement-learning-based dual-control methodology for complex nonlinear discrete-time systems with application to spark engine EGR operation. IEEE Trans Neural Network 2008;19(8):1369–88. https://doi.org/10.1109/TNN.2008.2000452.

[73] Shih P, Kaul BC, Jagannathan S, Drallmeier JA. Reinforcement-learning-based output-feedback control of nonstrict nonlinear discrete-time systems with application to engine emission control. IEEE Trans Syst Man Cybern Part B (Cybernetics) 2009;39(5):1162–79. https://doi.org/10.1109/TSMCB.2009.2013272.

[74] Maldonado BP, Kaul BC, Schuman CD, Young SR, Mitchell JP. Dilute combustion control using spiking neural networks. In: SAE technical paper 2021-01-0534. SAE International; 2021. https://doi.org/10.4271/2021-01-0534.

[75] Schuman CD, Potok TE, Patton RM, Birdwell JD, Dean ME, Rose GS, Plank JS. A survey of neuromorphic computing and neural networks in hardware. arXiv preprint arXiv:1705.06963 2017.

[76] Mitchell JP, Schuman CD, Potok TE. A small, low cost event-driven architecture for spiking neural networks on FPGAs. In: International conference on neuromorphic systems 2020. Association for Computing Machinery; 2020. https://doi.org/10.1145/3407197.3407216.

[77] Schuman CD, Mitchell JP, Patton RM, Potok TE, Plank JS. Evolutionary optimization for neuromorphic systems. In: Proceedings of the neuro-inspired computational elements workshop. Association for Computing Machinery; 2020. https://doi.org/10.1145/3381755.3381758.

[78] Schuman CD, Young SR, Mitchell JP, Johnston JT, Rose D, Maldonado BP, Kaul BC. Low size, weight, and power neuromorphic computing to improve combustion engine efficiency. In: 2020 11th international green and sustainable computing workshops (IGSC); 2020. p. 1–8. https://doi.org/10.1109/IGSC51522.2020.9291228.

[79] Luo Y, Maldonado B, Liu S, Solbrig C, Adair D, Stefanopoulou A. Portable in-cylinder pressure measurement and signal processing system for real-time combustion analysis and engine control. SAE Int J Adv Curr Prac Mobil 2020;2(6):3432–41. https://doi.org/10.4271/2020-01-1144.
[80] Jatana GS, Kaul BC. Determination of SI combustion sensitivity to fuel perturbations as a cyclic control input for highly dilute operation. SAE Int J Eng 2017;10(3):1011–8. https://doi.org/10.4271/2017-01-0681.
[81] Maldonado BP, Li N, Kolmanovsky I, Stefanopoulou AG. Learning reference governor for cycle-to-cycle combustion control with misfire avoidance in spark-ignition engines at high exhaust gas recirculation–diluted conditions. Int J Eng Res 2020;21(10):1819–34. https://doi.org/10.1177/1468087420929109.
[82] Garone E, Di Cairano S, Kolmanovsky I. Reference and command governors for systems with constraints: a survey on theory and applications. Automatica 2017;75:306–28. https://doi.org/10.1016/j.automatica.2016.08.013.
[83] Jade S, Hellström E, Larimore J, Stefanopoulou AG, Jiang L. Reference governor for load control in a multicylinder recompression HCCI engine. IEEE Trans Contr Syst Technol 2014;22(4):1408–21. https://doi.org/10.1109/TCST.2013.2283275.
[84] Caton JA. Combustion phasing for maximum efficiency for conventional and high efficiency engines. Energy Convers Manag 2014;77:564–76. https://doi.org/10.1016/j.enconman.2013.09.060.
[85] Pipitone E. A Comparison between combustion phase indicators for optimal spark timing. J Eng Gas Turbines Power 2008;130(5). https://doi.org/10.1115/1.2939012.
[86] Zhu GG, Haskara I, Winkelman J. Closed-loop ignition timing control for SI engines using ionization current feedback. IEEE Trans Contr Syst Technol 2007;15(3):416–27. https://doi.org/10.1109/TCST.2007.894634.
[87] Zhu Q, Wang S, Prucka R, Prucka M, Dourra H. Model-based control-oriented combustion phasing feedback for fast CA50 estimation. SAE Int J Eng 2015;8(3):997–1004. https://doi.org/10.4271/2015-01-0868.
[88] Emiliano P. Spark ignition feedback control by means of combustion phase indicators on steady and transient operation. J Dyn Syst Meas Contr 2014;136(5). https://doi.org/10.1115/1.4026966. 0 051021–051021–10.
[89] Xiao B, Wang S, Prucka RG. A semi-physical artificial neural network for feed forward ignition timing control of multi-fuel SI engines. In: SAE technical paper 2013-01-0324. SAE International; 2013. https://doi.org/10.4271/2013-01-0324.
[90] Zhang Y, Shen T. Cylinder pressure based combustion phase optimization and control in spark-ignited engines. Control Theory Technol 2017;15(2):83–91. https://doi.org/10.1007/s11768-017-6175-1.
[91] Maldonado BP, Zaseck K, Kitagawa E, Stefanopoulou AG. Closed-loop control of combustion initiation and combustion duration. IEEE Trans Contr Syst Technol 2020;28(3):936–50. https://doi.org/10.1109/TCST.2019.2898849.
[92] Stefanopoulou AG, Kolmanovsky I, Freudenberg JS. Control of variable geometry turbocharged diesel engines for reduced emissions. IEEE Trans Contr Syst Technol 2000;8(4):733–45. https://doi.org/10.1109/87.852917.
[93] Maldonado BP, Solbrig CE, Stefanopoulou AG. Feasibility and calibration considerations for selection of combustion control features. In: 2019 IEEE conference on control technology and applications (CCTA); 2019. p. 412–7. https://doi.org/10.1109/CCTA.2019.8920631.
[94] Johnson R, Kaczynski D, Zeng W, Warey A, Grover R, Keum S. Prediction of combustion phasing using deep convolutional neural networks. In: SAE technical paper 2020-01-0292. SAE International; 2020. https://doi.org/10.4271/2020-01-0292.

[95] Maldonado BP, Bieniek M, Hoard J, Stefanopoulou AG, Fulton B, Van Nieuwstadt M. Modelling and estimation of combustion variability for fast light-off of diesel aftertreatment. Int J Powertrains 2020;9(1−2):98−121. https://doi.org/10.1504/IJPT.2020.108423.

[96] Liu K, Li N, Rizzo D, Garone E, Kolmanovsky I, Girard A. Model-free learning to avoid constraint violations: an explicit reference governor approach. In: 2019 American control conference (ACC); 2019. p. 934−40. https://doi.org/10.23919/ACC.2019.8814772.

[97] Nicotra MM, Garone E. The explicit reference governor: a general framework for the closed-form control of constrained nonlinear systems. IEEE Contr Syst Mag 2018;38(4):89−107. https://doi.org/10.1109/MCS.2018.2830081.

[98] Daw CS, Edwards KD, Wagner RM, Green Jr JB. Modeling cyclic variability in spark-assisted HCCI. J Eng Gas Turbines Power 2008;130(5). https://doi.org/10.1115/1.2906176.

[99] Sen AK, Litak G, Edwards KD, Finney CEA, Daw CS, Wagner RM. Characteristics of cyclic heat release variability in the transition from spark ignition to HCCI in a gasoline engine. Appl Energy 2011;88:1649−55. https://doi.org/10.1016/j.apenergy.2010.11.040.

CHAPTER 9

Using deep learning to diagnose preignition in turbocharged spark-ignited engines

Eshan Singh, Nursulu Kuzhagaliyeva and S. Mani Sarathy

Clean Combustion Research Center (CCRC), King Abdullah University of Science and Technology, Thuwal, Western Province, Saudi Arabia

1. Introduction

In just the last decade, there has been an immense growth in application of soft computing techniques in internal combustion engine research. These include artificial neural networks, fuzzy-based approach, adaptive neuro-fuzzy interference system, gene expression programming, genetic algorithm, and particle swarm optimization [1]. Various models have been used in applications ranging from control and optimization to fault detection and performance/emission prediction. Increasingly, such techniques are replacing costly engine experiments and computationally expensive computational fluid dynamics (CFD) simulations [2]. They have been employed in conventional (spark and compression ignition engines) and alternate engine technologies (like Homogeneous Charge Compression Ignition (HCCI) [3] or Gasoline Compression Ignition (GCI) engines [4]) using both conventional and alternate fuels [5]. A review of work done in various areas of spark-ignition (SI) engine research is provided, laying out scope of future research in application of artificial intelligence (AI) in SI engines. Thereafter, we discuss one such application, diagnosing preignition in turbocharged spark ignited engines.

The existing literature can be broadly classified in a few categories, which often overlap and link with one another:

1.1 Fault detection

Common SI engine faults include abnormal combustion events, like knock or excessively high in-cylinder peak pressure. These are deleterious to the hardware on the long term. Several sensors' inputs have been used to detect faults using machine learning (ML). These include, but are not limited to, auditory signals from engine, crankshaft acceleration, vibration sensors (accelerometers), or pressure sensors. A fuzzy expert system, suggested by Kilagiz et al., could be used to detect faults in ignition, fuel system, intake and exhaust valves, based on fuel consumption and emission values [6]. Similar method was used by Wu et al., where they also compared a conventional back-propagation network with the proposed generalized regression neural network [7]. Shatnawi used the audio signals from the engines and extension neural network to diagnose faults [8]. Misfires are another class of abnormal combustion investigated by several researchers. Using simulations to train Multi-Layer Perceptron networks (MLP) feed-forward neural network and experimental data collected on a Toyota 4-cylinder engine to test the model, the system could detect misfire, along with information on its severity and location [9]. Using inputs from crank acceleration, Chen et al. used a regression-based ANN and an expanded neural network approach to detect misfires in a skip-fired engine [10]. More recently, preignition has become an annoyance in turbocharged SI engines. The rarity and stochastic nature of the abnormal combustion event makes it more difficult to predict. A few researchers have used ML algorithms to diagnose and detect preignition events as well [11,12].

1.2 Optimization and control

In a large fraction of existing literature, researchers have used data from in-cylinder pressure sensor (engine load, combustion phasing), temperature and pressure sensors (at the intake and exhaust side), engine speed and coolant temperature, etc., to optimize the engine operation for maximum power or efficiency. Simpler AI algorithms can replace the electronic controls as complexities in modern cars keep increasing with the increasing number of on-board computations. Use of ML techniques require lesser expertise and can be implemented easily with the rich data at disposal. Saraswati et al. combined neural networks with fuzzy logic-based control system for spark advance control to maintain peak pressure at 16 CAD aTDC. Artificial neural networks were used to give the peak pressure location, which was provided to fuzzy logic as an input to vary the spark timing. The system performed well, even in presence of external disturbances [13]. Lee et al. controlled fuel flow rate using fuzzy control system, to reduce emissions [14]. Another area of research is the use of AI to optimize the real-time engine:motor operation for hybrid vehicles for maximum efficiency or battery longevity (for plug-in hybrid electric vehicles) [15].

1.3 Predicting combustion parameters (phasing and cycle-to-cycle variation) and emissions

Combustion phasing is a control variable for all engine types. Predicting combustion phasing based on engine speed, load, intake air, coolant temperature, etc., have been undertaken by several authors. However, combustion phasing is a relatively easier variable to predict, in comparison to cycle-to-cycle variation (CCV). In-cylinder flow information (before ignition) has been used to predict CCV [16]. These inputs come from optical engine studies or, as is mostly the case, from CFD studies. Increasingly, AI models are replacing more expensive CFD techniques for predicting combustion parameters. Computation times have been reduced to less than half [2]. AI has also been used extensively to predict (and hence reduce) the emissions over the complete load-speed range of an engine. These include NO_x, nanoscale particulate matters, unburnt hydrocarbon, and carbon monoxide emissions [17].

2. Preignition detection using machine learning algorithm

While there is an extensive literature on application of artificial intelligence to various aspects of SI engines, there is a large scope of potential applications. These include, but are not limited to, more engine calibration and optimization, studying the K factor and applicability of high (or low) octane sensitivity fuels over the load-speed map, optimizing the blend ratios of various low-carbon alternative fuels, etc. Binary classification problems have been undertaken for detecting misfires, segregating high indicated mean effective pressure (IMEP) and CCV from low ones [18,19]. Similar methodology can be used to diagnose preignition cycles from normal combustion cycle.

Preignition is one of the bottlenecks to further downsizing, and efficiency improvements in spark-ignition engines. It refers to a stochastic occurrence of high peak pressures and pressure oscillations that can potentially destroy engine hardware. The occurrence is both rare and stochastic, making controlled experiments very difficult, eluding researchers from gaining any fundamental insight about the source of preignition [20]. However, several years of research have narrowed the possible sources to either deposit or lubricant initiated flame propagation [21,22]. The need for reduction in parasitic losses push for continued decrease in engine size and fewer cylinders. As the trend continues, engine operates at high load condition more often. Preignition events are more frequent at such operating conditions [23–25]. In the wake of such projections, it becomes imperative to diagnose preignition using relatively inexpensive sensors available on-board a vehicle.

Laboratory experiments often operate the engine at conditions increasing preignition events, there by allowing a large dataset to analyse. Commercial engines are modified for laboratory scale by disengaging all-but-one cylinders usually and employing various sensors and controls to operate the engine. While an on-road vehicle may use a lambda sensor and accelerometer, a laboratory engine will use a plethora of pressure (both high- and low-resolution sensors) and temperature sensors. On-road vehicle testing has shown that preignition in real-world driving condition may occur at higher speeds ($\sim$3000 rpm) more than at steady state laboratory conditions (the frequency of occurrence peaks around 2000 rpm in steady state experiments) [26]. Moreover, frequency response from preignition cycle has suggested that large amplitude can be observed in non-audible high frequency ranges (compared to conventional knock) [23].

Several methods have been employed previously to detect and diagnose preignition in spark-ignition engines. Use of in-cylinder pressure sensor is commonplace at laboratory scale. They provide reliable, high resolution pressure information, which can be used to classify a combustion cycle as normal or preignition cycle, based on start of combustion (from derivative of pressure data) or peak pressure or pressure oscillations (known as knock intensity). The pressure sensors used for preignition experiments are usually tolerant to even higher peak pressures ($\sim$300 bar) compared to the ones used in normal spark-ignited engine experiments. These sensors are extremely costly, hindering any scaled-up application in on-road vehicles (an exception can be SkyActiv-X engine from Mazda [27]).

Use of accelerometers is common in on-road vehicles. The accelerometers pick up engine vibrations in the knock frequency range (3–30 kHz), thereby allowing feedback control of spark timing. The fixed position of the accelerometer in vehicles does not allow cylinder specific information on pressure oscillations. Moreover, as observed by Singh et al. [23], the pressure oscillations in a preignition event may not follow the same frequency doublings as conventional knock. In such case, on-board accelerometer-based knock sensor may give erroneous information on preignition diagnosis.

Ion current-based sensors are in use at laboratory scale since late 19th century [28,29]. For preignition detection, Melby recorded ion-current signals to detect flame propagation initiated by a hotspot in 1953 [30]. Use of ion current sensor has been investigated by several researchers ever since, mostly as an inexpensive substitute to pressure sensors in on-road vehicles. Recently, Tong et al. [31,32] have developed such sensors for preignition detection for vehicles. However, their instability over a wide range of load and speed has inhibited their application in vehicles. Instead of conventionally used direct current source, alternating current source for ion current sensor have also been proposed to detect preignition more quickly [33]. Once again, vehicle-level implementation is unavailable for such technologies.

Amidst such technological advancements, data-driven technologies are making huge progress. Artificial intelligence and deep learning algorithms are being applied in various spheres of engine research. Engine calibration, fault detection, CCV, and engine-out emissions have been studied using ML methods by several researchers. High predictability and accuracy have been achieved by taking advantage of rich amount of data collected in engine experiments. Most in-cylinder fault detection studies have focused on diagnosing misfiring cycles, which occur due to incomplete or no combustion of fuel-air mixture. As opposed to other engine applications, fault detection presents added challenge of sparse data, due to a lower frequency of positives (misfiring or preigniting cycles).

ML is a field of artificial intelligence which uses training data to accomplish certain results from the machine without giving explicit set of instructions. Deep learning, as used in the current work, is a subfield of ML, aiming to replicate human brain operations in processing data, which may be unstructured and unlabeled. It is characterized by the effort to create computational models at several levels to learn representations of data with multiple levels of abstraction [34,35]. Different network architectures used for deep learning applications are discussed in the following section, along with other subparts employed in the two models used in the current study.

2.1 Feed forward multilayer neural networks

The most-often used network consists of an input layer and an output layer between several hidden layers. Each layer is fully connected (FC) to the previous and the following layer. The nodes between each layer are connected in the form of an acyclic graph, as shown in Fig. 9.1.

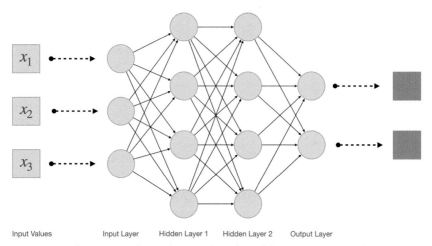

Figure 9.1 Fully connected neural network. *(Adapted from Patterson, J. Gibson, A. Deep learning: a practitioner's approach. O'Reilly Media, Inc.; 2007.)*

2.2 Convolutional neural networks

Convolutional neural networks (CNNs) are employed in case of high dimensional input features, like images, videos, or time-series data. A CNN learns data via progressive abstractions, whereby the learning progresses from one layer to another. In contrast to FC networks, neurons in one layer are connected to only a small number of neurons in previous layer. CNN is so named because the filtering operation performed by a feature map is mathematically defined as discrete convolution. The convolution operation is depicted in Fig. 9.2; it is the feature detector of CNNs. Input to

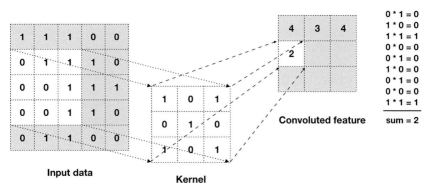

Figure 9.2 Convolutional operations. *(Adapted from Patterson, J. Gibson, A. Deep learning: a practitioner's approach. O'Reilly Media, Inc.; 2007.)*

convolution might be raw data or convoluted output from another convolution: it receives input, applies a convolution kernel, and generates features maps as the output. The convolutional kernel acts as a filter, allowing only certain types of information to pass from the input. As it can be seen, the kernel window slides across the input data range to generate the output matrix, eventually leading to one value as the output (4, in this case). The kernel matrix can be tuned to achieve varying output [35].

2.3 Recurrent neural networks

Recurrent neural networks (RNNs) are feed-forward neural networks that focus on modeling in the temporal domain. The distinctive feature of RNNs is their ability to send information over time steps. In their structure, RNNs have an additional parameter matrix for connections between time steps that promotes training in the temporal domain and exploitation of the sequential nature of the input. RNNs are trained to generate output where the predictions at each time step are based on current input and information from the previous time steps. RNNs are applicable to analysis of input in the time series domain. Data in this domain are ordered and context-sensitive, while elements in one timestep are related to elements in the previous time steps [37]. A comparison between normal and RNN neural network is shown in Fig. 9.3.

As previously mentioned, RNNs introduce the idea of recurrent connections. This type of wiring reconnects the output of a neuron in the hidden layer as a feed stream to the same hidden layer neuron. The recurrent connections and flow of information through the RNN can be visualized in Fig. 9.4.

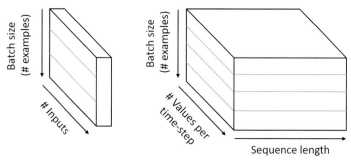

Figure 9.3 Comparison between normal and recurrent neural networks input vector. *(Adapted from Patterson, J. Gibson, A. Deep learning: a practitioner's approach. O'Reilly Media, Inc.; 2007.)*

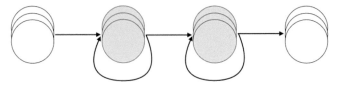

Figure 9.4 Feed forward flow for recurrent neural networks. *(Adapted from Patterson, J. Gibson, A. Deep learning: a practitioner's approach. O'Reilly Media, Inc.; 2007.)*

Fig. 9.5 demonstrates the term unrolling for time steps in an RNN. At each time step, every neuron in the hidden layer receives both input activation from the current input vector and from the previous hidden states. Thus, the output of the current time step is affected by the previous input vectors.

Long short-term memory (LSTM) is an artificial RNN architecture, which avoids the issue of vanishing gradient in classic RNN cells (which occurs when training RNN using back-propagation) by allowing gradients to flow unchanged. LSTM consists of a memory cell, input and forget units.

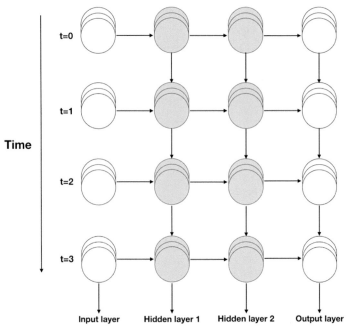

Figure 9.5 Unrolling for recurrent neural networks. *(Adapted from Patterson, J. Gibson, A. Deep learning: a practitioner's approach. O'Reilly Media, Inc.; 2007.)*

3. Activation functions

Activation functions define the output of a node given a particular input.

A rectified linear activation is used most often due to lower computational cost compared to sigmoid of tanh functions and because it overcomes the vanishing gradient problem. The rectified linear unit (ReLU), as it is known, is similar to a rectifier used elsewhere, wherein only the positive values are directly (or linearly) outputted, as is shown in Eq. (9.1). While the ReLU are significantly faster, they are prone to overfitting, when compared to sigmoid functions. This limitation can be overcome by using regularization techniques, such as dropouts [38].

$$f(x) = \max(0, x) \tag{9.1}$$

A softmax activation function is used in the output neural network layer to predict multinomial probability distribution. It simply converts the numbers into probabilities, based on the relative value of the numbers in each vector [37]. Hence, Softmax outputs one probability for each node (with value ranging between 0 and 1) and the sum of all outputs will be 1. Eq. (9.2) shows the mathematical description of a softmax function.

$$\sigma\left(\overrightarrow{z}\right)_i = \frac{e^{z_i}}{\sum_{j=1}^{K} e^{z_j}} \tag{9.2}$$

where σ = softmax, $\overrightarrow{z}$ = input vector, e^{z_i} = standard exponential function for input vector, K = number of classes in the multiclass classifier, and e^{z_j} = standard exponential function for output vector. As an example, the outputs of rectified linear and activation functions with an input vector $\overrightarrow{z}$ that contains values in the range -10 to 10 with a stepsize of 0.5 are demonstrated in Fig. 9.6.

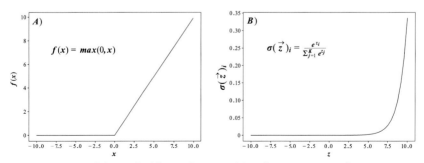

Figure 9.6 (A) Rectified linear function, (B) Softmax activation function.

4. Experiments and data extraction

Difficulty in reproducing preignition event at will has pushed the community into studying preignition in a statistical and probabilistic way. This has required a very large amount of dataset and thousands of engine cycles for any significant conclusions on the impact of operating conditions on preignition frequency. While such experiments are very costly and time-consuming, they offer a large and rich dataset to apply ML techniques on.

The engine experiments were run on a specially hardened AVL engine using a Euro V RON 95 Coryton gasoline. Details of the fuel, relevant to preignition, are shown in Table 9.1. Fig. 9.7 also shows the distillation

Table 9.1 Properties of the Coryton gasoline used in this study.

Research octane number (RON)	97.5
Motor octane number (MON)	86.6
Specific gravity (SG)	0.7485
Lower heating value (MJ/kg)	42.4
Energy density (MJ/L)	31.7
Aromatics (% v/v)	30.5
Olefins (% v/v)	8.2
Ethanol (% v/v)	5.0
H/C ratio	1.776
O/C ratio	0.015

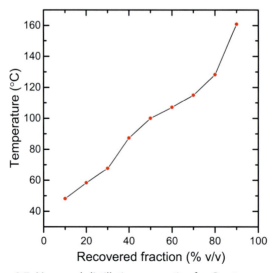

Figure 9.7 Measured distillation properties for Coryton gasoline.

curve for the fuel, in accordance with ASTM D-86 standard. An SAE 5E30 compliant lubricant was used in this study.

The test bed controls the engine speed at 2000 rpm. The gasoline is metered using a Coriolis meter and conditioned at 25°C before entering a reciprocating pumping unit, which increases the pressure to 150 bar, used for injecting gasoline directly in the chamber at −300 CAD aTDC. Various sensors are installed to monitor in-cylinder, intake and exhaust temperature and pressure. A lambda sensor is located 10 cm downstream of exhaust valve. In-cylinder pressure sensor is a high resolution AVL GU22CK capable of handling 300 bar peak pressures, while low resolution pressure sensors are mounted in intake and exhaust side. The response time of temperature sensor is ∼1 s (very low resolution).

To have diverse boundary conditions to train the model over, a wide range of exhaust back-pressures were used, which led to varying degree of preignition frequency. Intake air pressure was fixed for all the experiments at 1.9 bar. To achieve the fixed intake air pressure, intake temperature was decreased to compensate for increase in exhaust back pressure. The operating conditions can be summarized, as shown in Fig. 9.8. Such varying operating conditions provide a rich input dataset, while a huge number of cycles collected at each operating point (∼15,000 cycles) provide a huge experimental data set [22].

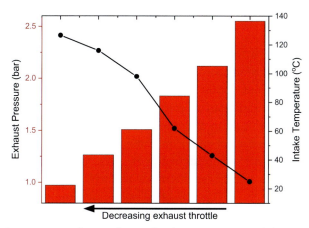

Figure 9.8 Operating conditions chosen for the training and validating the machine learning models.

5. Machine learning methodology

To extract the important parameters, the data needs preprocessing. The data from AVL engine is exported in an industry standard AVL file format. The format can be read using the opensource cross-platform general public license (GPL) tool for the analysis of internal combustion engine data, known as Catool. Data from all sensors are then exported in separate CSV files using Catool, which are then processed using a separate in-house MATLAB code, which provides crank-angle and cycle-based data, along with calculations for preignition count using various statistical methods. The high frequency sensor data are used for crank-angle based information, such as lambda, intake, exhaust, and in-cylinder data. Such data are referred to as multivariate time series data in ML problems. The cycle-based values, such as IMEP, knock intensity, CA10, CA50, CA90, etc., are calculated from the in-cylinder pressure data.

Classification of a combustion cycle into a normal or a preignition cycle is the most critical decision for the study. Researchers have used various markers to delineate preignition cycles. These include knock intensity (or pressure oscillations), start of combustion, peak pressure, or a combination of these. For the current study, start of combustion (CA05) is chosen as the marker. A robust statistics methodology is used according to the formula:

$$PI(CA05) \leq \theta_s \tag{9.3}$$

$$\theta_s = \overline{CA05}_S - 4.7 \times \sigma_S(CA05) \tag{9.4}$$

$$PI(CA05) \leq \theta_R \tag{9.5}$$

$$\theta_R = \overline{CA05}_R - 4.7 \times \sigma_R(CA05) \tag{9.6}$$

where θ_S and θ_R correspond to standard and robust cut-off values, $\overline{CA05}_S$ and $\overline{CA05}_R$ refer to standard and robust mean, $\sigma_S(CA05)$ and $\sigma_R(CA05)$ denote standard and robust standard deviation. Preignition events are marked with very early start of combustion (or very high knock intensity) which may skew the mean and standard deviation toward the preignition cycle. This impacts the cut-off value leading to fewer preignition cycles being counted, which is demonstrated in Fig. 9.9. Using robust statistics, the outliers have lesser impact on the mean and standard deviation leading to better preignition count. The classified cycles are then used to train, validate, and test the ML models.

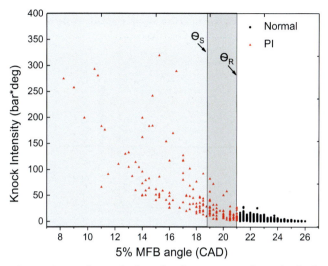

Figure 9.9 Comparison of standard and robust statistical methods for defining preignitions.

The in-cylinder pressure sensor can easily detect a preignition cycle due to invariably high peak pressure or earlier rise of pressure (from motoring curve) of a preignition cycle. However, the cost of these sensors prohibits using them in vehicles. Hence, the current work will focus on output from lambda sensors and low-resolution exhaust pressure sensors. Time-resolved data from lambda sensor and exhaust pressure sensor is also shown in Fig. 9.10. Lambda continues to increase in the preignition cycle and exhaust pressure sees a slight dip in preignition cycle.

Once data have been extracted using MATLAB, the data analysis library Pandas, and numerical computation library Numpy, in Python, were used for preprocessing. The ML library, Sklearn, popular for data mining and analysis, was also utilized. Using the complete cycle data as input will lead to large-dimensionality. This is avoided by selecting only +125 to +225 CAD aTDC as the region of interest. The start of this domain roughly corresponds to the exhaust valve opening time, hence the data before +125 CAD aTDC maybe be ignored without significant loss in information. The region of interest is shown in Fig. 9.11.

Principal component analysis (PCA) was carried out next onto the selected data inputs. PCA uses eigenvalue and eigenvector decomposition of the covariance matrix to project high dimensional input into a lower dimensional space. In other words, this tool facilitates the removal of

226 Artificial Intelligence and Data Driven Optimization of Internal Combustion Engines

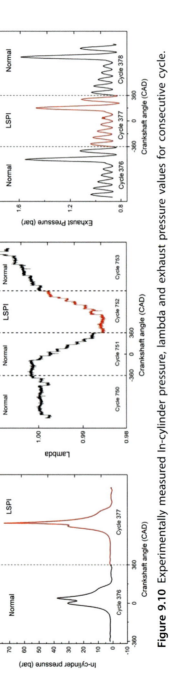

Figure 9.10 Experimentally measured In-cylinder pressure, lambda and exhaust pressure values for consecutive cycle.

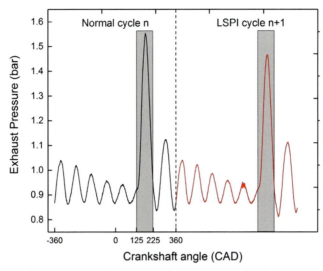

Figure 9.11 Chosen region of interest in the experimental exhaust pressure signal.

redundant dimensions and retains the most important (informative) ones. After interpolating data at 1 CAD apart, the 100 dimensions (125 CAD to 225 CAD) are reduced to 16 dimensions [39].

PCA for lambda is shown in Fig. 9.12. The projections of principal components show no demarcation between preignition cycles (red triangles) and normal cycles (hollow circles). The condition does not improve for three-dimensional (3D) projections of three principal components. Moreover, three principle components were enough to capture 99% of cumulative variance in the data.

Similarly, principle component analysis on exhaust pressure data yielded well demarkated class between preignition and normal cycles, as shown in Fig. 9.13. About 99% of the variance in the exhaust pressure data can be explained used five principal components or more. A 3D plot of projections of principal components shows even better class separation between preignition and normal cycles.

About 80% of data was used for training and validation. For the training part, the input and output were fed into the model to fit the weights. Thereafter, validation ensured that the model generalized the patterns learned during the training. These data were monitored during hyperparameter tuning, optimal number of layers, nodes, and learning rate. The validation set provided unbiased evaluation of a model on the training set. The test set was used to assess the generalization error. The data were

228 Artificial Intelligence and Data Driven Optimization of Internal Combustion Engines

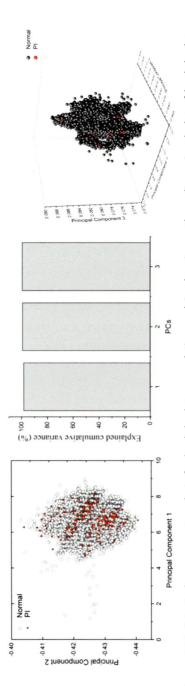

Figure 9.12 Principal component analysis for lambda data, cumulative explained variance, three-dimensional plot of lambda data projection on three principle components.

Using deep learning to diagnose preignition in turbocharged spark-ignited engines 229

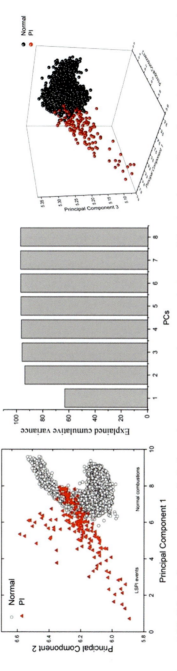

Figure 9.13 Principal component analysis for exhaust pressure data, cumulative explained variance, three-dimensional plot of exhaust pressure data projection on three principle components.

normalized using min-max scalar to accelerate the learning process. The data were then shuffled to further enhance the model generalization. The two-dimensional data were reshaped into 3D for further processing in a CNN, a process known as data mini batching.

Two different models were explored to investigate the ML capabilities of preignition data. In the first model (known as Model 1 here onwards), the PCA is used to reduce the dimensions of the data. As previously discussed, PCA extracts the critical information from a multidimensional feature, leading to faster computations. In the second model (Model 2), the time-series is provided to fit the model. This is done to check if any critical information is missing from the PCA analysis in lieu of faster computation. The two models are further discussed in more detail below:

6. Model 1: Input from principal component analysis

After investigating several network architectures (over 50 epochs), a combination of convolutional and recurrent neural layers, stacked linearly, was chosen. 3D data from two stacked cycles go through three convolutional layers with decreasing dimensions in each layer, with a kernel size of 2. Its output goes through a RNN with LSTM unit. Thereafter, the FC layer is added to the softmax activation function, which yields the output probabilities, as discussed previously.

A sequential model was build using Keras libraries and compiled using Tensorflow backend. As opposed to numerical predictions, the current predictions are binary classifications.

Hence a loss function, based on binary cross-entropy, was used. The expression is provided in Eq. (9.7).

$$\text{loss} = -t\log(p) - (1-t)\log(p) \tag{9.7}$$

An ADAM optimization algorithm with a default learning rate of 0.001 was used. Other details are provided in Table 9.2.

Table 9.2 Fixed training parameters for Model 1 and Model 2.

Optimization algorithm	ADAM
Learning rate	0.001
Loss function	Cross-entropy
Maximum number of epochs for training	1000
Validation split	0.2

Table 9.3 Hyperparameter optimization for Model 1.

No. of filters in CONV1D$_1$	[16, 32]
No. of filters in CONV1D$_2$	[32, 64]
No. of filters in CONV1D$_3$	[64, 128]
No. of hidden layers in LSTM$_1$	[32, 64, 128]
Dropout$_1$	[0.1, 0.2]
Dropout$_2$	[0.1, 0.2, 0.3]
Dropout$_3$	[0.1, 0.2, 0.3, 0.4]
Batch size (B)	[1000, 3000, 5000]

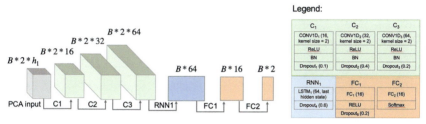

Figure 9.14 Final network architecture from hyperparameter tuning (Model 1).

The proposed model exhibits several hyperparameters that need to be chosen. Hyperparameter optimization was performed using exhaustive search approach, where each hyperparameter was randomly selected from an array of proposed values. Overall, 100 models with different hyperparameter settings were evaluated, the optimal model architecture was selected based on reported minimum validation loss during 1000 epochs. The list of hyperparameter values used for the manual search is given in Table 9.3.

The final network architecture with optimized hyperparameters for Model 1 is illustrated in Fig. 9.14. In order to avoid over-fitting, batch normalization is performed after each convolutional layer for dynamical normalization. Dropouts were also incorporated after each layer of the network architecture. Dropouts randomly remove a specified percentage of weights, thereby introducing noise into the neural network to force the model to learn to generalize well enough to deal with noise.

7. Model 2: Time series input

Time series data between +125 and +225 was input without PCA. An even higher dimensional input in this model meant increased number of convolutional layers and number of feature maps, which is followed by two

Table 9.4 Hyperparameter optimization for Model 2.

No. of filters in CONV1D$_4$	[128, 256]
No. of filters in CONV1D$_5$	[64, 128]
No. of filters in CONV1D$_6$	[32, 64]
No. of filters in CONV1D$_7$	[16, 32, 64]
No. of filters in CONV1D$_8$	[16, 32]
No. of hidden layers in LSTM$_2$	[16, 32, 64]
No. of hidden layers in LSTM$_3$	[16, 32, 64]
Dropout$_4$	[0.2, 0.3]
Dropout$_5$	[0.2, 0.3]
Dropout$_6$	[0.1, 0.2, 0.3]
Dropout$_7$	[0.1, 0.2, 0.3, 0.4]
Dropout$_8$	[0.1, 0.2, 0.3, 0.4]
Dropout$_9$	[0.1, 0.2]
Batch size (B)	[1000, 3000, 5000]

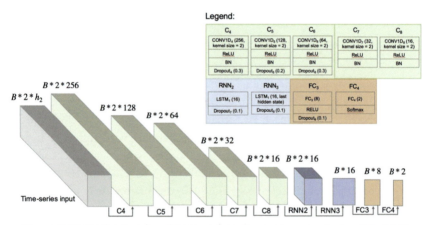

Figure 9.15 Final network architecture from hyperparameter tuning (Model 2).

layers of LSTM and FC layers. Thereafter softmax actional provides the output probabilities, just as in previous model. The parameters are shown in Table 9.4 and Model 2 architecture with selected hyperparamater setting is shown in Fig. 9.15.

8. Model metrics

Various metrics are used to gauge the performance of the models studied in the current work. These are shown in Eqs. (9.8)–(9.10).

$$\text{Precision} = \frac{\text{True Positives}}{\text{True Positives} + \text{False Positives}} \quad (9.8)$$

$$\text{Recall} = \frac{\text{True Positives}}{\text{True Positives} + \text{False Negatives}} \quad (9.9)$$

$$F1\ \text{score} = 2 * \frac{\text{Precision} * \text{Recall}}{\text{Precision} + \text{Recall}} \quad (9.10)$$

9. Results and discussion

9.1 Training and validation losses

Learning curves enable visualizing how well the machine is getting trained as the data are fed. After each epoch, in addition to training learning curve, the generalization error is assessed with the validation loss, which is evaluated on validation data (20% of total data set). In Fig. 9.16, both training and validation losses drastically decrease and reach minimum after 50 epochs. It is notable that lower training and validation losses are achieved during training of Model 1 in comparison to Model 2. This indicates that Model 1 with PCA input exhibits greater capability to distinguish PI cycles than Model 2 with the time-series input.

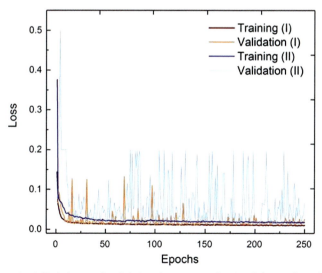

Figure 9.16 Training and validation loss curves for Model 1 and Model 2.

Table 9.5 Precision and Recall values for Model 1 and Model 2.

Model/Metric	Precision (%)	Recall (%)	F1 score (%)
Model 1	74.31	85.14	79
Model 2	67.7	84	75

The results for the unseen test set predictions are reported in terms of precision, recall, and F1 score values for two models. From Table 9.5, predictions with Model 1 demonstrate higher precision (74.31%) and recall (85.14%) values, which result in higher F1 score (79%). Based on three common metrics used for binary classification, Model 1 outperforms Model 2 and detects more preignition events with reasonably high precision.

10. Conclusions

Artificial intelligence has been extensively applied to SI engines over the last decade. They have been used for control and optimization of engines, predicting combustion parameters and emissions, and detecting faults in engines. Preignition is an abnormal combustion event, whose diagnosis was attempted in this work, using relatively inexpensive lambda sensor and low-resolution exhaust pressure sensor. Two different models were used in the current study to diagnose preignition cycle among normal cycles. One model uses a PCA analysis to extract critical information from the full cycle data from the two sensors, while the other model uses full data from the two sensors. The model using PCA performed better in achieving the best precision and recall simultaneously.

While the current study investigated, and successfully established, a data-driven diagnosis of preignition based on inexpensive lambda sensor and low resolution exhaust pressure sensors, several more sensors can be used to achieve even higher precision and recall rates. Researchers have already signaled a relation between other parameters (in-cylinder residual content, engine-out hydrocarbon emissions, and particulate matters) and preignition cycle [22,40,41]. Detecting stochastic and abnormal events will always be difficult, but preignition experiments provide a vast amount of rich data to make accurate predictions about their onset. Future studies will aim at predicting preignition events based on data available from on-board sensors and laboratory based sensors. Studies need to be undertaken at vehicle level and on-road tests to fully understand the efficacy of employing such models in real vehicles.

References

[1] Shrivastava N, Khan ZM. Application of soft computing in the field of internal combustion engines: a review. Arch Comput Methods Eng 2018;25(3):707—26.
[2] Kavuri C, Kokjohn SL. Exploring the potential of machine learning in reducing the computational time/expense and improving the reliability of engine optimization studies. Int J Engine Res 2020;21(7):1251—70.
[3] Vaughan A, Bohac SV. Real-time, adaptive machine learning for non-stationary, near chaotic gasoline engine combustion time series. Neural Netw 2015;70:18—26.
[4] Badra J, et al. Engine combustion system optimization using CFD and machine learning: a methodological approach. In: Internal combustion engine division fall technical conference. American Society of Mechanical Engineers; 2019.
[5] Wong KI, et al. Modeling and optimization of biodiesel engine performance using advanced machine learning methods. Energy 2013;55:519—28.
[6] Kilagiz Y, et al. A fuzzy diagnosis and advice system for optimization of emissions and fuel consumption. Expert Syst Appl 2005;28(2):305—11.
[7] Wu J-D, Liu C-H. An expert system for fault diagnosis in internal combustion engines using wavelet packet transform and neural network. Expert Syst Appl 2009;36(3):4278—86.
[8] Shatnawi Y, Al-Khassaweneh M. Fault diagnosis in internal combustion engines using extension neural network. IEEE Trans Ind Electron 2013;61(3):1434—43.
[9] Chen J, Randall RB. Improved automated diagnosis of misfire in internal combustion engines based on simulation models. Mech Syst Signal Process 2015;64:58—83.
[10] Chen SK, et al. Machine learning for misfire detection in a dynamic skip fire engine. SAE Int J Engines 2018;11(6):965—76.
[11] Wolf P, et al. Pre-ignition detection using deep neural networks: a step towards data-driven automotive diagnostics. In: 2018 21st international conference on intelligent transportation systems (ITSC). IEEE; 2018.
[12] Kuzhagaliyeva N, et al. Using deep neural networks to diagnose engine pre-ignition. Proceedings of the Combustion Institute; 2020.
[13] Saraswati S, Agarwal PK, Chand S. Neural networks and fuzzy logic-based spark advance control of SI engines. Expert Syst Appl 2011;38(6):6916—25.
[14] Lee S, Howlett R, Walters S. Emission reduction for a small gasoline engine using fuzzy control. IFAC Procs Vol 2004;37(22):185—90.
[15] Murphey YL, et al. Intelligent hybrid vehicle power control—Part I: machine learning of optimal vehicle power. IEEE Trans Veh Technol 2012;61(8):3519—30.
[16] Dreher D, et al. Deep feature learning of in-cylinder flow fields to analyze cycle-to-cycle variations in an SI engine. Int J Engine Res 2020. 1468087420974148.
[17] Pu Y-H, Reddy JK, Samuel S. Machine learning for nano-scale particulate matter distribution from gasoline direct injection engine. Appl Therm Eng 2017;125:336—45.
[18] Hanuschkin A, et al. Investigation of cycle-to-cycle variations in a spark-ignition engine based on a machine learning analysis of the early flame kernel. Proceedings of the Combustion Institute; 2020.
[19] Hanuschkin A, et al. Machine learning—based analysis of in-cylinder flow fields to predict combustion engine performance. Int J Engine Res 2021;22(1):257—72.
[20] Singh E. Mechanism triggering pre-ignition events and ideas to avoid and suppress pre-ignition in turbocharged spark-ignited engines. 2019.
[21] Singh E, et al. Effect of mixture formation and injection strategies on stochastic pre-ignition. SAE Technical Paper; 2018.
[22] Singh E, Dibble R. Mechanism triggering pre-ignition in turbo-charged engines. SAE Technical Paper; 2019.

[23] Singh E, Dibble R. Effectiveness of fuel enrichment on knock suppression in a gasoline spark-ignited engine. SAE Technical Paper; 2018.
[24] Singh E, Dibble R. Knock, auto-ignition and pre-ignition tendency of fuels for advanced combustion engines (FACE) with ethanol blends and similar RON. SAE Technical Paper; 2020.
[25] Singh E, et al. Effect of different fluids on injection strategies to suppress pre-ignition. SAE Technical Paper; 2019.
[26] Michlberger A, et al. On-road monitoring of low speed pre-ignition. SAE Technical Paper; 2018.
[27] Hitomi M, et al. Control device for spark-ignition engine. Google Patents; 2005.
[28] Stoddard EJ. Electrical igniter for gas-engines. Google Patents; 1902.
[29] Mcmullin FR. Electric sparker for gas-engines. Google Patents; 1901.
[30] Melby A, Diggs D, Sturgis B. An investigation of preignition in engines. SAE Technical Paper; 1954.
[31] Tong S, et al. Cycle resolved combustion and pre-ignition diagnostic employing ion current in a PFI boosted SI engine. SAE Technical Paper; 2015.
[32] Tong S, et al. Knock and pre-ignition detection using ion current signal on a boosted gasoline engine. SAE Technical Paper; 2017.
[33] Wilstermann H, et al. Ignition system integrated AC ion current sensing for robust and reliable online engine control. SAE Technical Paper; 2000.
[34] Goodfellow I, et al. Deep learning, vol. 1. Cambridge: MIT press; 2016.
[35] LeCun Y, Bengio Y, Hinton G. Deep learning. Nature 2015;521(7553):436—44.
[36] Patterson J, Gibson A. Deep learning: a practitioner's approach. O'Reilly Media, Inc.; 2017.
[37] Gulli A, Pal S. Deep learning with Keras. Packt Publishing Ltd; 2017.
[38] Gal Y, Ghahramani Z. Dropout as a bayesian approximation: representing model uncertainty in deep learning. In: International conference on machine learning; 2016.
[39] Wold S, Esbensen K, Geladi P. Principal component analysis. Chemometr Intell Lab Syst 1987;2(1—3):37—52.
[40] Costanzo VS, et al. Fuel & lubricant effects on stochastic preignition. SAE Int J Adv Curr Pract Mobil 2019;1:259—77. 2019—01-0038.
[41] Haenel P, et al. Influence of ethanol blends on low speed pre-ignition in turbocharged, direct-injection gasoline engines. SAE Int J Fuels Lubr 2017;10:95—105. 2017—01-0687.

Further reading

[1] Cay Y. Prediction of a gasoline engine performance with artificial neural network. Fuel 2013;111:324—31.
[2] Gölcü M, et al. Artificial neural-network based modeling of variable valve-timing in a spark-ignition engine. Appl Energy 2005;81(2):187—97.
[3] Yücesu HS, et al. Comparative study of mathematical and experimental analysis of spark ignition engine performance used ethanol—gasoline blend fuel. Appl Therm Eng 2007;27(2—3):358—68.
[4] Kiani MKD, et al. Application of artificial neural networks for the prediction of performance and exhaust emissions in SI engine using ethanol-gasoline blends. Energy 2010;35(1):65—9.
[5] Najafi G, et al. Performance and exhaust emissions of a gasoline engine with ethanol blended gasoline fuels using artificial neural network. Appl Energy 2009;86(5):630—9.
[6] Najafi G, et al. SVM and ANFIS for prediction of performance and exhaust emissions of a SI engine with gasoline—ethanol blended fuels. Appl Therm Eng 2016;95:186—203.

[7] Kapusuz M, Ozcan H, Yamin JA. Research of performance on a spark ignition engine fueled by alcohol–gasoline blends using artificial neural networks. Appl Therm Eng 2015;91:525–34.
[8] Cay Y, et al. Prediction of engine performance for an alternative fuel using artificial neural network. Appl Therm Eng 2012;37:217–25.
[9] Danaiah P, Ravi Kumar P, Rao Y. Performance and emission prediction of a tert butyl alcohol gasoline blended spark-ignition engine using artificial neural networks. Int J Ambient Energy 2015;36(1):31–9.
[10] Isin O, Uzunsoy E. Predicting the exhaust emissions of a spark ignition engine using adaptive neuro-fuzzy inference system. Arabian J Sci Eng 2013;38(12):3485–93.
[11] Togun NK, Baysec S. Prediction of torque and specific fuel consumption of a gasoline engine by using artificial neural networks. Appl Energy 2010;87(1):349–55.
[12] Tasdemir S, et al. Artificial neural network and fuzzy expert system comparison for prediction of performance and emission parameters on a gasoline engine. Expert Syst Appl 2011;38(11):13912–23.

Index

Note: 'Page numbers followed by "f" indicate figures and "t" indicate tables.'

A

Acceleration strategies, 75–76
Accelerometers, 216
Activation functions, 221
Active learning loop, 153–154
Active optimizer (ActivO), 12
  basic algorithm, 160–161, 161f
  computational fluid dynamics (CFD)
    engine optimization, 165f, 171–179, 172t, 175f
    simulations, 159
  convergence criteria, 163–164
  design optimization, 159
  dynamic exploration, 164–165
  exploitation, 164–165
  flowchart, 165f
  genetic algorithm using derivatives (GENOUD), 167
  indicated specific fuel consumption (ISFC), 172–173
  internal combustion (IC) engines, 159
  µGA, 173–174, 177f
  optimum design parameters, 176t
  particle swarm optimization (PSO), 167
  query strategies, 161–163, 163f
  robustness, 168–169
  simulation-driven design optimization (SDDO), 159
  two-dimensional cosine mixture function, 165–171, 168f
Adaptive mesh refinement (AMR), 72, 74–75, 75f
Aromatics, 37
Artificial Intelligence (AI), 2, 2f, 49–52, 126–127
  black-box, 186–187
  coefficient of variation (CoV), 188–189
  combustion stability control, neural networks for, 193–198
  artificial neural networks, 193–195, 203t
  spiking neural networks, 195–198, 196f–197f
  cumulative distribution function (CDF), 191
  cycle-to-cycle variability (CCV), 188–193
  dilute combustion instability, 187–189
  exhaust gas recirculation (EGR), 185–186
  fuel-engine co-optimization, 58
    combustion events mitigation, 15–16
    fuel formulation, 13–15
    internal combustion engine (ICE), 4–13, 6f, 13f
  fuel property prediction, machine learning models, 54, 55t
  haystack needle, 52–53
  high throughput screening, 52–53
  homogeneous charge compression ignition (HCCI), 185–186
  indicated mean effected pressure (IMEP), 188–189
  kernel density estimator (KDE), 191
  model-free dilute limit identification, learning reference governor for, 199–204, 200f
    avoiding misfire events, learning reference governor for, 194t, 202–204
    constrained combustion phasing control problem, 199–202, 201f
  nuclear magnetic resonance (NMR), 54, 56f
  physics-based (white-box) approach, 190
  proportional-integral-derivative (PID)-style reactive approaches, 185
  reaction discovery, 57
  spark-ignition, 189–193

239

Artificial Intelligence (AI) (*Continued*)
  three-way catalyst (TWC), 188
Artificial neural networks (ANNs), 8−15, 51, 142
Automated machine learning-genetic algorithm (AutoML-GA), 141−154
  active learning loop, 153−154
  artificial neural networks (ANNs), 142
  extreme gradient boosting (XGB), 142
  hyperparameter selection, 142−145
  Kernel ridge regression (KRR), 141
  Nu support vector regression, 141
  optimal hyperparameter values, 154
  problem setup, 145−146
  regularized polynomial regression (RPR), 141

B

Black-box optimization algorithms, 35−36, 186−187
Bottom Dead Center (BDC) temperatures, 29−30
Brake-specific fuel consumption (BSFC), 8−12

C

Coefficient of variation (CoV), 188−189
Combustion noise level (CNL), 8−12
Combustion parameters, 27, 215
Combustion stability control, neural networks for, 193−198
  artificial neural networks, 193−195, 203t
  spiking neural networks, 195−198, 196f−197f
Compression ignition (CI) engine, 73−74
Computational efficiency, 119
Computational fluid dynamic (CFD) model, 5, 57, 74−76, 104−105
  adaptive mesh refinement, 74−75, 75f
  detailed chemistry acceleration strategies, 75−76
  engine optimization, 165f, 171−179, 172t, 175f
  model setup and validation, 113, 114t
  simulations, 159
  validation, 118−119
Constrained optimization formulation, 32−33
Convergence acceleration, 91−96, 93f−94f, 97f
  genetic algorithms, 79−81, 80f
  multiobjective framework, 84−88, 85f, 87f
  pioneering investigations, 81−84
Convergence criteria, 163−164
Convolutional neural networks (CNNs), 218−219, 218f
Cumulative distribution function (CDF), 191
Cycle-to-cycle variability (CCV), 188−193

D

Data fitting, 117−118
Data science, 48
Deep learning (DL), 2, 2f, 47
  gradient-based optimization, 13−14
Design of experiments (DoE), 73, 125−126
  computational efficiency, 119
  computational fluid dynamic (CFD) model, 5−7, 104−105
  model setup and validation, 113, 114t
  validation, 118−119
  data fitting, 117−118
  design space construction, 107−109
  efficient geometry, 119
  engine and fuel specifications, 113
  evaluation method, 116
  full engine operation range, 119
  gasoline compression ignition (GCI), 104−105
  mesh manipulation, 119
  methodologies, 106−113
  model-based design optimization, 112−113
  objective variables, 116
  optimization, 117−119
  response surface model (RSM), 104, 109−112

sample size, 119
test engine specifications, 114t
Design optimization, 159
Design space construction, 107—109
Dilute combustion instability, 187—189
Direct injection spark ignition (DISI) combustion system, 5—7
Dynamic exploration, 164—165

E

Efficient geometry, 119
Electronic control unit (ECU), 8—12
Engine geometry generation, 76—78
   method of forces, 77—78, 78f
   method of splines, 76—77, 77f
Engine optimization
   acceleration strategies, 74—79
      computational fluid dynamics (CFD), 74—76
      engine geometry generation, 76—78
      virtual injection model, 78—79, 79f
   adaptive mesh refinement (AMR), 72
   compression ignition (CI) engine, 73—74
   design of experiments (DoEs), 73
   internal combustion engine (ICE), 71
   modeling framework, 74—79
   optimization methods
      convergence acceleration, 91—96, 93f—94f, 97f
      genetic algorithms, 79—81, 80f
      multiobjective framework, 84—88, 85f, 87f
      pioneering investigations, 81—84
   response surface method (RSM), 73
Exhaust aftertreatment system, 58
Exhaust gas recirculation (EGR), 127, 185—186
Exploitation, 164—165
Extreme gradient boosting (XGB), 142

F

Fault detection, 213—214
Feed forward multilayer neural networks, 217
Fuel-engine co-optimization, 58

Fuel formulation workflow, 48—49
Fuel performance metrics, 28—31
Fuel property prediction, machine learning models, 54, 55t
Fuel representation, 47—48
Full engine operation range, 119

G

Gasoline compression ignition (GCI), 5—7, 104—105
Gaussian process—based surrogate model optimization algorithm, 35—37, 36f
Gaussian process (GP) models, 28
Generative adversarial networks (GANs), 13—14, 53
Genetic algorithm (GA), 28, 33—34, 34f
   computational fluid dynamics (CFD), 7—8
Genetic algorithm using derivatives (GENOUD), 167
Genetic programming (GP), 12
Global search methods, 144

H

Haystack needle, 52—53
High throughput screening, 52—53
Homogeneous charge compression ignition (HCCI), 185—186
Hyperparameter selection, 142—145
   global search methods, 144
   line or grid search, 143
   manual selection, 142—143
   random search, 143
   selecting hyperparameters, automated strategies for, 143—145

I

Ignition delay time (IDT), 30
Indicated mean effected pressure (IMEP), 188—189
Indicated specific fuel consumption (ISFC), 172—173
Industrial revolution, 1, 1f
Internal combustion engine (ICE), 71, 159
   Artificial Intelligence (AI), 126—127

Internal combustion engine (ICE) (*Continued*)
   fuel-engine co-optimization, 4–13, 6f, 13f
   automated machine learning-genetic algorithm, 141–154
     active learning loop, 153–154
     artificial neural networks (ANNs), 142
     extreme gradient boosting (XGB), 142
     hyperparameter selection, 142–145
     Kernel ridge regression (KRR), 141
     Nu support vector regression, 141
     optimal hyperparameter values, 154
     problem setup, 145–146
     regularized polynomial regression (RPR), 141
   data examination, 129–132
   design of experiments (DoE), 125–126
   exhaust gas recirculation (EGR), 127
   machine learning-genetic algorithm (ML-GA) approach, 126–127, 132–140
     optimization methodology, 132–134
     postprocessing, 139–140
     repeatability of, 134–139, 135t, 137f
     robustness, 139–140
     variable domain extension, 136–139, 138t
   training, 129–132
Isoparaffins, 37

K

Kernel density estimator (KDE), 191
Kernel ridge regression (KRR), 141
K-means clustering method, 50

L

Long short-term memory (LSTM), 52, 220

M

Machine learning (ML), 2, 2f, 8–13, 49
   accelerometers, 216
   activation functions, 221
   algorithm, 215–220
   combustion parameters, 215
   convolutional neural networks (CNNs), 218–219, 218f
   data extraction, 222–223
   fault detection, 213–214
   feed forward multilayer neural networks, 217
   internal combustion engines (ICEs), 12–13, 13f
   long short-term memory (LSTM), 220
   methodology, 224–230, 226f
   model metrics, 232–233
   optimization and control, 214
   preignition, 215–216
     detection, 215–220
   principal component analysis (PCA), 225–227, 230–231, 230t–231t
   recurrent neural networks (RNNs), 219–220, 219f
   styles, 3–4, 3f
   time series input, 231–232
   training and validation losses, 233–234
Machine learning-genetic algorithm (ML-GA) approach, 126–127, 132–140
   optimization methodology, 132–134
   postprocessing, 139–140
   repeatability of, 134–139, 135t, 137f
   robustness, 139–140
   variable domain extension, 136–139, 138t
Manual selection, 142–143
Mesh manipulation, 119
Mixed-mode combustion, 28–31
Model-based design optimization, 112–113
Model-free dilute limit identification, learning reference governor for, 199–204, 200f
   avoiding misfire events, learning reference governor for, 194t, 202–204
   constrained combustion phasing control problem, 199–202, 201f
Model metrics, 232–233
Multimode operation, 27

N

Naphthenes, 37
Neural network (NN), 28
Nondominated sorting genetic algorithm (NSGA-II), 8–12

Nonlinearity, 50
Nuclear magnetic resonance (NMR), 54, 56f
Nu support vector regression, 141

O

Objective variables, 116
Olefins, 37
One-dimensional (1D) models, 28
Optimal hyperparameter values, 154
Optimization, 117–119
    constrained optimization formulation, 32–33
    formulation, 41–42
    Gaussian process–based surrogate model optimization algorithm, 35–37, 36f
    genetic algorithm, 33–34, 34f
Optimum design parameters, 176t

P

Paraffins, 37
Particle swarm optimization (PSO), 167
Physics-based (white-box) approach, 190
Potential energy surface (PES), 57
Preignition, 215–216
    detection, 215–220
Principal component analysis (PCA), 225–227, 230–231, 230t–231t
Proportional-integral-derivative (PID)-style reactive approaches, 185

Q

Query strategies, 161–163, 163f

R

Radial basis function (RBF), 8–12
Random search, 143
Reaction discovery, 57
Recurrent neural networks (RNNs), 13–16, 219–220, 219f
Regularized polynomial regression (RPR), 141
Reinforcement learning, 3–4, 3f
Response surface method (RSM), 73, 104, 109–112
Robustness, 168–169
Root-mean-square-errors (RMSE), 31

S

Simplified molecular-input line-entry system (SMILES), 13–14
Simulation-driven design optimization (SDDO), 159
Spark-ignition (SI), 27, 189–193
Supervised learning, 3–4, 3f
Support vector machine (SVM), 47

T

Test engine specifications, 114t
Three-dimensional computational fluid dynamics (3D-CFD) simulation models, 27
Three-way catalyst (TWC), 188
Time series input, 231–232
Traditional chemical space screening, 53
Transportation fuels, 47–52
    Artificial Intelligence modeling approaches, 49–52
    fuel formulation workflow, 48–49
    fuel representation, 47–48
Two-dimensional cosine mixture function, 165–171, 168f

U

Unsupervised learning, 3–4, 3f

V

Virtual injection model, 78–79, 79f

Z

Zero-dimensional (0D) models, 28
Zero-RK software package, 29

Printed in the United States
by Baker & Taylor Publisher Services